多样性的中国草原 DIVERSE RANGELANDS OF CHINA

一滴水
生态摄影
丛书

A Single Drop of Water
Ecological Photography
Album Series

多样性的中国草原

DIVERSE RANGELANDS OF CHINA

陈建伟 著/摄
CHEN JIANWEI

中国林业出版社
CFPH China Forestry Publishing House

图书在版编目（CIP）数据

多样性的中国草原 / 陈建伟著、摄. -- 北京 : 中国林业出版社, 2025. 5. -- (一滴水生态摄影丛书).
ISBN 978-7-5219-2787-0
Ⅰ. S812.5
中国国家版本馆CIP数据核字第2024ZX2944号

审图号：GS京（2025）0832号

本书由北京林业大学草业与草原学院院长董世魁教授主审。

中国林业出版社·自然保护分社（国家公园分社）
责任编辑：袁丽莉 肖静
特约编辑：田红
科学审核：董世魁
英文翻译：柴晚锁 黄晓凤 张曼玉
英文审校：柴晚锁
图片编辑：崔林
装帧设计：崔林

出版发行：中国林业出版社（100009#北京市西城区刘海胡同7号）
http://www.cfph.net
电话：（010）83143577
E-mail：forestryxj@126.com
印刷：北京雅昌艺术印刷有限公司
版次：2025年5月第1版
印次：2025年5月第1次
开本：787mm × 1092mm 1/12
印张：20
字数：300千字
定价：320.00元

谨把此书：献给全球首次联合国“国际草原与牧民年（IYRP 2026）”！

This book is dedicated to the United Nations’ first “International Year of Rangelands and Pastoralists (IYRP) 2026”!

雪山下的草原•
青海玛沁
Rangelands under the snow-capped mountain — Maqin, Qinghai

阿尼玛卿山是黄河源头最大的山脉，为昆仑山支脉，和西藏的冈仁波齐、云南的梅里雪山和青海玉树的尕朵觉沃被并称为藏传佛教四大神山，主峰玛卿岗日海拔6282米（画面中的尖山峰）。在永久冰雪带以下的高山冰缘带，分布着两种特殊的草原生态系统——高山流石滩稀疏植被生态系统和高山垫状植被生态系统。

The Amnye Machen Range, a branch of the Kunlun Mountains and the largest one situated at the source of the Yellow River, ranks among the 4 sacred mountains of the Tibetan Buddhism together with Mount Kailash in Tibet, Meili Snow Mountain in Yunnan and Mount Gardo Jowo in Yushu, Qinghai. What is seen in the picture is the 6283-meter peak of this grand range, namely the Maqing Gangri. Lying just below the permanent snow and ice zone are 2 types of very special rangeland ecosystems, the sparsely-vegetated alpine scree ecosystem and the alpine cushion vegetation ecosystem.

野花似锦树如剑•
新疆博尔塔拉

A satin blanket of wild flowers and sky-piercing swords of trees — Bortala, Xinjiang

夏日的夏尔西里，漫山遍野的草原上鲜花烂漫，伟岸挺拔的云杉剑一样直指蓝天，一幅美丽的森林草原的天然画卷呈现在我们面前。这里由于极少受到人类活动干扰，自然资源基本处于原始状态，被称为新疆的“最后一片净土”和“西北不可多得的天然基因库”。

At the summer time, the rolling mountains of Sharhili is covered with a thick colorful blanket of wild flowers dotted occasionally with towering spruces that point sharply into the blue sky, presenting us a picturesque view of the untainted forest rangelands. Almost untrodden by human footprints, the natural resources here are basically kept at a primeval state and is therefore hailed as the “last pure land” of Xinjiang as well as the “rare natural genome pool of North-western China”.

冬营盘•
内蒙古锡林郭勒
Winter Grazing Lands — Xilingol, Inner Mongolia

在广袤的内蒙古高原上，分布着非常优质的典型草原生态系统。数千年来，这里孕育了丰富深厚的草原文化、游牧文明，其影响曾经横跨了整个欧亚大陆。典型草原特殊的生存环境塑造了牧民随季节变换、驱赶牛羊转换草场的千年习俗，每年牧民都要转场于季节草场之间。图为蒙古牧民冬天放牧生活的“冬营盘”。

High quality typical rangelands ecosystem can be widely seen on the vast plateau of Inner Mongolia, where unique rangeland culture and nomadic civilization have been in existence for thousands of years and in turn projected their influences over the entire Eurasian continent. The unique environment of the vast rangelands has given rise to the time-honored tradition of moving between different grazing lands with the evolution of seasons that is strictly observed by local folks. The pastoralists would often move regularly between the different seasonal pastures. What is shown in the picture is the winter grazing lands of the Mongolian herdsmen.

A 雪山下的荒漠草地•新疆哈密

Desert rangelands below the snow-capped mountains — Hami, Xinjiang

在气候干燥寒冷的北疆，这里呈现的是一幅非常特殊的生态景观：远处的天山终年白雪皑皑，雪线下温带森林生长茂密，山下就是大片的荒漠草原，近处是干旱的沙地，多年生的猫头刺呈馒头状分布在沙砾中。别看猫头刺那个长相刺楞楞的，春季绵羊、山羊采食其花和小叶，骆驼采食其嫩枝，马甚至刨食其根。

What we encounter here in the cold and dry northern Xinjiang is a very special ecological view: an impressive mixture of landscape composed of the permanently snow-capped Tianshan Mountain in the distance, the verdant dense temperate forests below the snowline, the vast stretches of desert rangelands at the foot of the mountain, and the arid sandy lands close by that is covered thickly by perennial bun-shaped *Oxytropis aciphylla* Ledeb. Don't be cheated by the prohibitive look of this prickly plant, they constitute precious food for animals inhabiting here in the food-scarce spring times, with sheep and goat feeding on their blossoms and newly-folding leaves, camels on their tender twigs, and horses even on their deep-embedded roots.

美丽的草原我的家•甘肃玛曲
Home on the beautiful rangelands — Maqu, Gansu

玛曲的藏语意为“黄河”，在这片中华民族母亲河——黄河源区处，支流众多的黄河在近一万平方千米的玛曲草原上缓缓流过，静静地滋润着这片美丽富饶的草原，并在此形成黄河九曲中的第一曲。牛羊在这里享受着无忧无虑的生活，藏族群众在这美丽富饶的草原上世世代代繁衍生息。

The term *Maqu* in Tibetan language means “the Yellow River”. The region from where the mother river of China is originated, the 10,000m^2-Maqu rangelands makes up the home for numerous tributaries that meander quietly through this fertile and beautiful place. This is also the site where the Yellow River makes the first among its 9 awesome turns on its journey towards the sea. This rolling rangelands, in addition to providing a sanction where livestock graze leisurely and carefree, sustains generation after generation of Tibetan folks who take it as their homes.

序 言

FOREWORD

我很高兴看到《多样性的中国草原》一书出版。本书是《多样性的中国森林》《多样性的中国湿地》《多样性的中国荒漠》的姊妹篇，是作者在视野广阔、知识深厚基础上的又一力作。

在了解本书的科学背景后，我非常高兴应邀为这本《多样性的中国草原》作序。我想借此机会为读者指出本书的几个突出特色和创新。

一、本书科学、简要地讲述了草原生态系统是“草地—畜群—人类”相互影响、相互制约、动态平衡的生态系统，而人为的调控是草原生态系统发生变化的决定性因素。现在草原的大面积退化、沙化就是人工调控不当、动态平衡被人为打破的结果。草原生态系统是一个与森林、湿地、荒漠、农田、海洋都不相同的、极为特殊的生态系统，我们应该秉承“师法自然”的理念，科学地保护和可持续利用好草原。

二、本书讲述了草原和森林、草原和湿地、草原和荒漠生态系统之间的辩证关系，强调各类生态系统之间的过渡、交融，以及你中有我、我中有你的自然属性。本书特别强调了各生态系统之间可能的转换，例如，荒漠和草原之间，由于全球气候变暖，人努力、天帮忙，完全可能向良好方面转化。本书还特别强调了不应该僵化地对待退化草原的观点是值得肯定的，这种观点对于我们修复退化草原、正确理解和贯彻“山水林田湖草沙”综合治理的理念是大有裨益的。

三、本书针对草原生态系统的特殊性，一改前面对于森林、湿地、荒漠生态系统的呈现和表述中尽量采用原始的、自然的、没有人类干扰破坏痕迹的图片的常规做法，而是大量采用了有人、有畜的草原生产和生活的图片，让读者正确理解草原生态系统中“草地—畜群—人类”三者相互依存、同为生命共同体的特殊内涵。草原生态系统具有“生态、生产、生计”三大功能，是与其他生态系统都不一样的极为特殊而复杂的生态系统。

四、本书首次应用生态摄影的新理念、新方式来全面讲述中国草原的精彩生态故事。生态摄影强调的是思想性、科学性和艺术性的统一，以生态价值观为准则，从而达到全新的境界。用生态摄影的手段来讲好中国草原的故事，需要图文并茂，既要给人草原各种景观的直观感受，又要让人理解其中的科学内涵；既要全面兼顾，又要重点突出；既要肯定草原保护和修复取得的成绩，又不能回避存在的问题。另外，书中还针对各大区不同生态系统的保护和修复提出建议和生态思考（如何可持续发展），这些都相当考验作者的功力。

五、为了让读者更好地了解草原生态系统中的野生动植物，本书最后还专门将书中涉及的野生动植物物种的中文名、学名，以及在《中国国家重点保护野生动植物名录》、《濒危野生动植物种国际贸易公约》附录和《世界自然保护联盟濒危物种红色名录》中的级别一一列出，便于读者查询和理解。

习近平总书记提出“科技创新、科学普及是实现创新发展的两翼”，这本《多样性的中国草原》采用了生态摄影的新理念，用精彩的图片和洗练的文字，给我们讲述了一个又一个生动的中国草原生态故事。本书不仅着重宣传了中国草原保护的成绩，还提出了当前存在的问题，同时融入了作者的深度思考，是一本难得的中国草原优秀科普图书。在当前国家政府机构职能转型中，对于草原工作的重视程度是过去从来没有过的，我们要充分宣传和发挥中国草原对于全国生态文明社会建设中的重大作用，既要充分发挥草原作为生态屏障的生态功能，又要可持续地保持好草原的特殊生产功能。本书的出版发行，也较好地贯彻和践行了这个理念。

我一生都在从事中国草原的科研教学工作，经历了中国草原的建设事业从零起步到目前蓬勃发展的全过程。当前草原资源被纳入“山水林田湖草沙”的综合治理之中，草原的科学研究和宣传教育得到了国家前所未有的高度重视，草原生态系统的保护和修复工作也提高到了国家前所未有的高度。我亲眼看到一代代青年俊秀都纷纷奔赴全国草原科研、教育、管理、生产的各层次、多方面的战线，为中国草原的保护和建设作出各自的贡献，感到十分欣慰。希望他们随着国家生态文明建设的发展，为中国的草原事业作出更大贡献！

中国工程院资深院士 任继周

2025年1月8日

序 言

FOREWORD

I am very delighted to see the debut of *Diverse Rangelands of China*, another landmark eco-album that the author has photographed and written on basis of his extensive exposure to the wilderness and rich knowledge assets, adding one more impressive work to the series that he had previously published, which includes *Diverse Forests of China, Diverse Wetlands of China, Diverse Deserts of China* and others.

It is a great pleasure for me to, particularly after learning about the rich and solid scientific foundation upon which it is based, write a foreword for *Diverse Rangelands of China*. I would like to take this opportunity to point out to would-be readers a few key innovative aspects in this book that are highly impressive to me.

1) This book illustrates in a brief yet scientifically-based manner that the rangeland ecosystem is a system characterized by the mutually influencing, mutually checking and delicately balanced relationship between grasses, livestock and people, in which human-initiated intervention and regulation plays a decisive role in shaping the development trajectories of rangeland ecosystems. The massive degradation and sandification of rangelands that has occurred over recent decades is just the consequence of ill-conceived human intervention and regulation that had negatively impacted on this delicate balance. Given that rangeland ecosystem is a very special one, different from the ones composed of forests, wetlands, deserts, oceans or agricultural croplands, a "go-after-nature" or "nature-based" philosophy must be adopted in our endeavor to protect and use our rangelands in science-based and sustainable ways.

2) This book brings into sharp relief the dialectical inter-ecosystem relation between rangelands and forests, between rangelands and wetlands, and between rangelands and deserts by highlighting the transitional, inter-exchangeable and mutually-involving properties of all these ecosystems. In particular, it's stressed in the book that these ecosystems are all liable to changes. Take for instance, thanks to the influence of global warming, if only human-made efforts and favorable natural climate conditions are properly made use of, it is totally possible to make real a positive conversion between deserts and rangelands. The author also emphasizes in the book that it is a highly commendable proposal that the degraded rangelands should be viewed in a less rigid and indoctrinated way, since this proposal will be very conducive and beneficial to us in our endeavor to restore the degraded rangelands, to correctly understand and earnestly implement the comprehensive vision that calls for integrated and synthetic approaches in the management of mountains, rivers, forests, croplands, lakes, rangelands and sandy lands.

3) With regard to the peculiarity of the rangeland ecosystem, the author has boldly deviated from conventional approaches that typically present and depict desirable forest, wetland and desert ecosystems through images that focus intensely on their primitive, natural, and human-impact-free features. Instead, photos that zoom in on the daily life and productive activities of rangeland dwelling folks — with both people and livestock obviously present — are abundantly included, so as to drive home to readers the idea that views the rangeland ecosystem as the shared community where grasses, livestock and people exist side by side in mutually dependent and mutually complementing ways. Unlike any other type of its counterparts, the rangeland ecosystem is a very special and complex one that simultaneously plays triple roles in the provision of ecological, productive and livelihood functions.

4) This book marks the first attempt to tell about the excellent eco-stories featuring China's rangelands through the new and innovative media of eco-photography, which calls for a seamless blending of thoughts, science and artistic tastes and is guided by ecological values so as to bring this art into new realms. Telling convincing and impressive Chinese rangeland stories through eco-photographing is a very demanding job. It takes eye-appealing pictures as well as good command of languages to present readers with compelling visual images of rangeland landscape on one hand, and easily accessible knowledge about the underlying science on the other hand. It needs to be holistic and comprehensive yet have clear focuses; it needs to be commendatory about the achievements that have been made so far while not

covering up problems that still exist in our rangeland protection and restoration efforts. Practical suggestions pertaining to the protection and restoration of various ecosystems in each of the major regions, together with thoughtful eco-reflections (concerning how to make sure these efforts are sustainable), are also presented in the book. All of these testifies convincingly to the great language finesse and artistic tastes of the author.

5) In order to help readers better understand the wild flora and fauna species living in the rangeland ecosystem, the author has attached at the end a thorough appendix — giving a complete list of all wildlife species covered in the book, including their Chinese and Latin names, their respective status on the *List of Key Protected Wildlife Species of China*, the *CITES* species appendices, and the *IUCN Red List of Threatened Species* — which the readers can easily consult whenever necessary.

As how President Xi Jinping puts it, "Scientific innovation and extensive outreach of science are the two wings for realizing innovation-powered development." *Diverse Rangelands of China* resorts to the latest eco-photographing concepts, telling pictures as well as concise yet compelling languages to tell us, in vivid and lifelike manners, one after another of China's moving stories related with its rangelands. In addition to highlighting the great achievements that the country has attained in rangeland conservation and protection, the author also makes penetrating observations with regard to problems that still exist in the field and put forward some fairly insightful suggestions for their solution. All these makes this book an invaluable resource that readers interested in learning more about rangelands in China can turn to for some detailed yet easily accessible information. At this highly favorable moment when the Chinese government is determined to re-configurate the responsibilities and functions of its administrative authorities, and when unprecedented importance is being attached to work related to rangelands, it's imperative for us to take this advantageous opportunity to stress on the important roles that rangelands play in driving the nation's ecological progresses further forwards and make the fullest use of such roles. Sufficient attentions must be paid to rangelands so as to make sure that both their ecological functionsin safeguarding national ecological security and their unique productive functions in guaranteeing sustainable livelihood can be used to maximum degrees. The publication of this book marks a key milestone in our endeavor to put this guiding principle into earnest practices.

As a person who has dedicated almost my entire academic career to researches and teaching in this field, I have personally witnessed China's unswerving efforts in the conservation and protection of its rangelands, turning this vison from a mere promise in its early stages to the notable prominence that we see today. Deemed now as an integral part of the mountains-rivers-forests-croplands-lakes-rangelands-and-sandy lands synthetic management approach, rangeland resources are now valued to an extent that they have never seen before, as evidenced by the unprecedented importance that the country currently attaches to the scientific researches, educational campaign and ecosystem conservation/restoration with regard to rangelands. I feel very much delighted and inspired to see generation after generation of talented young men and women getting enthusiastically involved in rangeland researches and education, and in rangeland productive and managerial programsas well, where each of them is contributing their due shares. I'm convinced that, as China strives forward in its ecological progress, they will make still grander contributions to the country's ambitions pertaining to rangeland sustainability.

REN Jizhou, Senior Chinese Academy of Engineering (CAE) Academician

Jan. 8th, 2025

目 录
CONTENTS

序 言 17

Foreword

导 言 22

Introduction

第一章　内蒙古高原草原区 34

Chapter 1 The Inner Mongolian Plateau Rangeland Region

1-1. 森林草地生态系统 Forest grassland ecosystem
1-2. 典型草原生态系统 Typical rangeland ecosystem
1-3. 沙化草原生态系统 Sandified rangeland ecosystem
1-4. 草甸生态系统 Meadow ecosystem
1-5. 可持续发展 Sustainable development

第二章　西北山地盆地草原区 72

Chapter 2 The Mountainous and Basin Northwestern Rangeland Region

2-1. 森林草地生态系统 Forest grassland ecosystem
2-2. 干旱荒漠草原生态系统 Dry desert rangeland ecosystem
2-3. 极干旱荒漠草原生态系统 Extremely dry desert rangeland ecosystem
2-4. 草甸生态系统 Meadow ecosystem
2-5. 可持续发展 Sustainable development

第三章　青藏高原草原区 110

Chapter 3 The Qinghai-Tibetan Plateau Rangeland Region

3-1. 高山流石滩稀疏植被生态系统 Sparsely-vegetated alpine scree ecosystem

3-2. 高寒垫状植被生态系统 Alpine cushion vegetation ecosystem

3-3. 高寒草甸生态系统 Alpine meadow ecosystem

3-4. 高寒草原生态系统 Alpine rangeland ecosystem

3-5. 高寒荒漠草原生态系统 Alpine desert rangeland ecosystem

3-6. 森林草地生态系统 Forest grassland ecosystem

3-7. 可持续发展 Sustainable development

第四章　东北华北平原山地丘陵草原区 176

Chapter 4 The Northern and Northeastern Plain and Hilly Rangeland Region

4-1. 森林草地生态系统 Forest grassland ecosystem

4-2. 暖性草丛生态系统和暖性灌草丛生态系统 Warm tussock ecosystem and warm shrubbery and tussock ecosystem

4-3. 草甸生态系统 Meadow ecosystem

4-4. 可持续发展 Sustainable development

第五章　南方山地丘陵草原区 204

Chapter 5 The Southern Hilly Rangeland Region

5-1. 热性草丛生态系统和热性灌草丛生态系统 Hot tussock ecosystem and hot shrubbery and tussock ecosystem

5-2. 干热稀树草地生态系统 Savannah-like sparse-tree grassland ecosystem

5-3. 草甸生态系统 Meadow ecosystem

5-4. 可持续发展 Sustainable development

物种名录 227

List of species covered in the book

后　记 232

Postscript

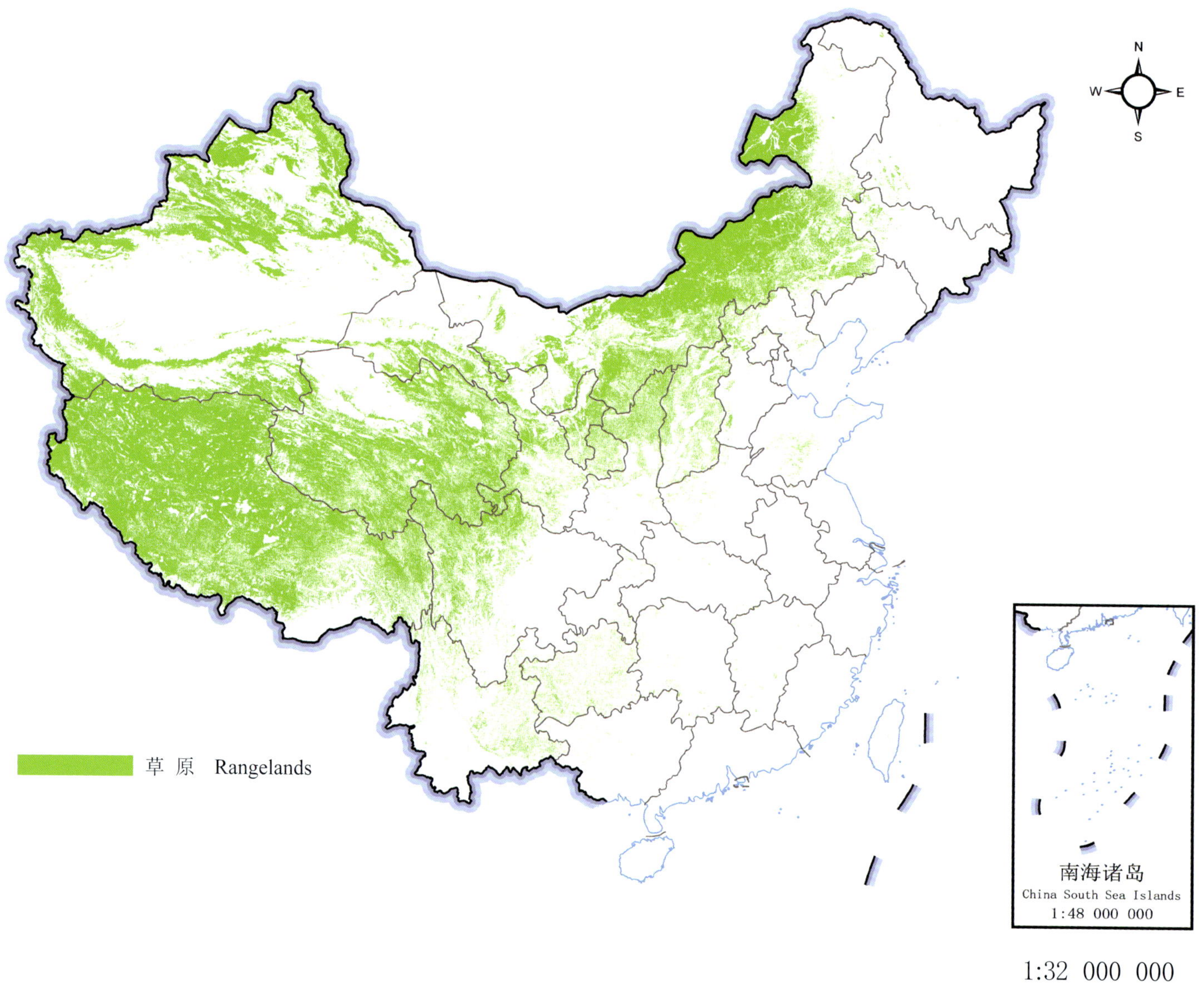

中国草原分布图

Sketch-map of rangelands in China

CHAPTER 1 第一章

内蒙古高原
草原区

The Inner Mongolian Plateau Rangeland Region

羊马成群伴河曲•
内蒙古东乌珠穆沁旗
Sheep and horses grazing on river banks — Dongwuzhumuqin Banner, Inner Mongolia

乌拉盖草原是世界上保存最完好的天然草原之一，以典型草原为主，乌拉盖河蜿蜒流过，河畔马、羊成群。这里不仅是肥美草原的代名词，还因是《狼图腾》故事的发生地及电影的拍摄地而闻名于世。这里曾经是内蒙古生产建设兵团农垦历史的见证地，被列入全国第一批“红色草原”名单中。

Wulagai Grassland, through which the Wulagai River meanders quietly, is one of the best-preserved natural rangelands dominated by typical grasslands in the world. Herds of cattle, horses and sheep graze leisurely on banks of the river. This place has not only become a synonymy of "fertile grassland", but also gained a fame for being the site where the highly-acclaimed movie *Wolf Totem* was shot. It once witnessed the history of the Inner Mongolia Production and Construction Corps, and is therefore listed among the first batch of "Red Grasslands" of the country in honor of this glorious tradition.

I
II
III
IV
V
北京★
Beijing
南海诸岛
China South Sea Islands
1:64 000 000
图例 Legends
I 内蒙古高原草原区
The Inner Mongolian Plateau Rangeland Region
II 西北山地盆地草原区
The Mountainous and Basin Northwestern Rangeland Region
III 青藏高原草原区
The Qinghai-Tibetan Plateau Rangeland Region
IV 东北华北平原山地丘陵草原区
The Northern and Northeastern Plain and Hilly Rangeland Region
V 南方山地丘陵草原区
The Southern Hilly Rangeland Region
国界 National Border
1∶32 000 000

该区北部东界为大兴安岭西坡，北到黑龙江国界，往南包括科尔沁沙地，经燕山、太行山，向西南抬升到黄土高原、鄂尔多斯高原，主体部分包括大兴安岭、呼伦贝尔高原、锡林郭勒高原、科尔沁沙地、乌兰察布高原、鄂尔多斯高原、河套平原、黄土高原、大青山、阴山及贺兰山以东等地区，是世界著名的欧亚大草原的东端。该区草原属温带半湿润半干旱草原生态系统。

太平洋暖湿气流由东南向西北逐渐减弱，降水随之递减，而热量变化的总趋势是从东北到西南逐渐递增，但其增加的趋势又因海拔的升高而减弱。因此，草原植被类型呈有规律的变化，即森林草原—典型草原—荒漠草原的逐步过渡。在大兴安岭西部、南部和大青山等，其基部为森林草地生态系统和典型草原生态系统。在西部的贺兰山，其基部为沙化草原生态系统。

此外，本区还分布山地草甸、沼泽草甸、盐生草甸等多种草甸类型。

The northern part of this region borders in the east on the western slope of Daxinganling and extends northwards until the Heilongjiang River, which marks the borderline of the country. The southern part passes through the Horqin Sandy Land, the Yanshan Mountains, the Taihang Mountains until it reaches the elevated Loess Plateau and Ordos Plateau in the southwest. It mainly consists of the Daxinganling, Hulunbeier Plateau, Xilingol Plateau, Horqin Sandy Lands, Ulaanchabu Plateau, Ordos Plateau, Hetao Plain, Loess Plateau, Daqingshan Mountains, Yinshan Mountains, and areas east to the Helan Mountains. This region marks the eastern end of the world-famous Eurasian steppe, with temperate semi-humid, semi-arid rangelands ecosystem being the prevalent type.

The warm and humid airflow from the Pacific Ocean weakens gradually as it moves from the southeast to the northwest, leading at the same time to ever declining precipitation. While the heat in this region demonstrates an ever increasing trend as we move along the northeast to southwest direction, this overall increasing trend is weakened as the altitude rises. For this reason, the vegetations in this region change in an obvious pattern, namely, a gradual transition from forest rangelands to typical rangeland and finally to desert rangelands. At the western and southern foot of Daxinganling and that of Daqingshan Mountains, the dominant types are forest grassland ecosystem and typical rangeland ecosystem. At the foot of Helan Mountains situated in the western part of this region, the dominant type is desert rangeland ecosystem.

Typical meadows, marshy meadows and saline meadows are also widely found in this region.

1-1 森林草地生态系统

FOREST GRASSLAND ECOSYSTEM

在该区的东部，与森林生态系统相邻的西侧，有一条狭长的森林草地生态系统带，这是森林生态系统向典型草原生态系统过渡的一种生态系统类型，主要分布于森林向草原的过渡地区，也分布于典型草原地带山地的阴坡与丘顶，同时也见于某些山地草原带森林的上部。

森林草地生态系统是本区草原生态系统中最湿润、生态条件最好的一种类型，降水量较高，年平均降水量450 mm 左右，处于高原面上及高原面隆起的山地上，植物种类组成比较丰富。

这里是最好的宜牧区，往往也被认为是宜农地、宜林地。而其中的很大一部分地区的土壤基质是沙土，大量的放牧、割草、开垦农地、植树等活动，使这里土地沙化，土地生产力急剧下降，甚至丧失，很多草原生态系统严重退化，荒漠化严重。这不仅使畜牧业遭受损失，开垦的农田、林地也不能长期维持，我们今天看到的分布在这里的斑斑块块的大片沙地，就是这类教训的明证。

In the eastern part of the region that borders on the western edge of adjacent forest ecosystem, there exists a narrow belt of forest grasslands ecosystem, a type marked by the gradual shift from forest ecosystem to typical rangeland ecosystem and often found at the transitional areas between forests and rangelands. This type of ecosystem also exists on shady slopes and hilltops of typical rangeland areas, or at the upper part of forests in certain mountainous rangeland areas.

Forest grassland ecosystem is a type that is the most humid and has the most favorable ecological conditions among all the rangeland ecosystems found in the region. It usually comes with relatively high precipitation, approximately 450 mm in annual average, and is often located on plateau surfaces or mountains rising from plateau surfaces. Plants typically come in a rich diversity.

This type of ecosystem makes for the best grazing lands, but meanwhile is also often considered as suitable for agriculture and forestry development. The soil matrix in a large part of this region is sandy, making it highly vulnerable to activities like extensive grazing, grass-cutting, reclamation of agricultural land and tree planting. As a result of sandification caused by these activities, the productivity of lands here tends to decline sharply, or to disappear altogether, leading in consequence to serious degradation and desertification of many grassland ecosystems. This has not only caused losses to the husbandry industry, but also made it impossible to maintain the reclaimed farmland and forest land for long. The vast patches of sandy areas that we see here today are clear evidence of what might happen when this type of ecosystem degrades.

生态景观组合图·内蒙古阿尔山
A combination of ecological landscapes — Arsan, Inner Mongolia

地处大兴安岭西南山麓的阿尔山为中低山地貌，这里森林、草原、河谷、流水相互依偎，又浑然一体。从高空看，大地满被绿色覆盖。但当这些森林、山地草甸、沼泽草甸、河滩、河流等反映在阳坡、阴坡及阳光下、云影中时，却呈现出了各种深浅不同的绿色，仅这一隅似乎就用尽了大地画卷中所有的绿色颜料。

Located in the southwestern foothills of the Daxing anling Mountains, Arshan is a terrain composed of low and medium hills, upon which forests, grasslands, river valleys and flowing water are nestled together in a harmonious manner. When viewed from the air, the entire place would look as if covered in a uniformed blanket of green. However, when each of these — forests, grassy slopes, swampy meadows, riverbanks, rivers — is viewed from different perspectives or against varied backgrounds like sunny slopes, shady slopes, bright sunshine, or cloud shadow, the green will appear in a rich variety of dark and light shades, as if all the different combinations of colors had been used to decorate up this tiny corner of the earth.

西辽河上游源区•
内蒙古赛罕乌拉
The source of upper reach of the West Liao River — SaihanWula, Inner Mongolia

赛罕乌拉是契丹族的故土和发祥地，现在建有内蒙古赛罕乌拉国家级自然保护区，也是世界生物圈保护区。它是一个以保护珍稀濒危野生动植物及其赖以生存的森林、草原、湿地等多样生态系统为主的综合性自然保护区。图上这片森林草地是西辽河支流的查干沐沦河源头水源涵养区。

SaihanWula, the homeland and birthplace of the Khitan people, is now the site of the Inner Mongolia SaihanWula National Nature Reserve as well as part of World Biosphere Reserve. It is a comprehensive nature reserve that protects both the rare and endangered wild fauna and floral species inhabited here and the rich diversity of forest, rangeland and wetland ecosystems that these protected species rely on for survival. The forest grassland shown in the photo is the source of Chagan Mulun River, a major tributary of the West Liao River.

森林草地带•
河北塞罕坝
A belt of forest rangelands — Saihanba, Hebei

在内蒙古高原南缘，一条森林草地带镶嵌在燕山北麓。由于这里地处纬度较低的暖温带，以条带状、块状分布在草地上的森林，已经由以针叶林为主的林相变成以落叶阔叶林为主的林相。

At the southern edge of the Inner Mongolia Plateau, a belt of lush forest rangelands belt is tucked cozily at the northern foothills of the Yanshan Mountains. Situated in low-latitude warm temperate zone, the forest strips and blocks that dotted over the grasslands have changed from a forest phase dominated by coniferous stands to one where deciduous broad-leaved stands prevail.

稀树草原·内蒙古克什克腾旗
Rangelands dotted by sparse trees —
Keshketeng Banner, Inner Mongolia

内蒙古克什克腾旗地处大兴安岭南端山地和燕山余脉交汇地带，这里呈现着大片稀树草原景观。克什克腾旗内设有黄岗梁、青山、乌兰布统、桦木沟、白音敖包、达里诺尔等自然保护区，是我国县级行政区中自然保护区数量最多的县域之一。

Keshketeng Banner in Inner Mongolia is located at a place where the southern end of the Daxinganling Mountains converges with the ending section of the Yanshan Mountains. It presents to us a telling example of the impressive views of sparsely tree-dotted rangelands. Being the home of numerous nature reserves like Huanggangliang, Qingshan, Ulanbutun, Huamugou, Baiyin Ovoo and Dalinuoer, Keshketeng Banner ranks high among its counterparts across the country for the great number of nature reserves it has.

柳兰·内蒙古额尔古纳
Chamerion angustifolium —
Erguna, Inner Mongolia

柳兰为多年生草本植物，广泛分布于中国东北、华北地区，常成片生于林缘、林间隙地和森林草地中，性喜光，耐寒，适生于湿润肥沃、腐殖质丰富的土壤。

Chamerion angustifolium is a perennial herbaceous plant that is widely found to thrive in groves at forest edges, glades, or over forest rangelands in northeastern and northern China. Being sunlight-prone and cold-resistant, this plant grows well in fertile and humus-rich soil with high moisture-content.

草花与昆虫•河北坝上
Blossoms and insects —
Bashang, Hebei

每当夏季来临，林边草地上就盛开了各种各样的小花，这几枝名为麻花头的多年生草本植物正在开花，吸引来了众多的蚂蚁。呈现在你面前的这小小微观世界，其实正在讲述小草抓紧时间利用昆虫来繁育后代的故事。

When summer arrives each year, grasslands that stretch out from the edge of forests would be densely dotted with a huge variety of tiny flowers. The perennial *Klasea centauroides* your see in the photo are in full bloom, attracting a large number of ants. In spite of its miniature size, what this micro-world presents you is a story about how grasses take the fullest advantages of insects to help them to propagate.

求爱•
内蒙古白云敖包
Courtship —
Baiyun Ovoo, Inner Mongolia

黑琴鸡发情时在森林草地上有相对固定的场地，俗称“斗鸡盘”。雄鸟几只或十几只飞到草地上后开始鸣叫，尾羽垂直向上展开呈扇状，翅膀下垂，不时地跳起来与其他雄鸟搏斗。图中的两只雄性黑琴鸡低头哈腰、嘴里不停地咕咕叫着，向心仪的雌性求爱不舍，而这时雌性黑琴鸡昂头挺胸缓迈步，像一位骄傲的公主。

Black grouses (*Lyrurus tetrix*) inhabiting forest rangelands tend, during the breeding seasons, to have their fixed sites for courtship, which are commonly known as the “cockfighting ring”. A handful, sometimes, even up to a dozen, of male birds would show up on the grasslands and start to chirp. Wings drooping, they would spread their tail feathers out in vertical fan-like shapes as they occasionally jump up to fight with rivalling courters. What you see in the picture are two male black grouses that are, with bowed heads and incessant cooing, competing eagerly to win the heart of the female, who at this moment is parading proudly nearby like a princess, with her head held up high.

森林草地斜阳•
河北围场
Slanting sunlight over forest rangelands —
Weichang, Hebei

围场县御道口牧场地处内蒙古高原东南缘，处于阴山山脉东端、大兴安岭山脉南端、燕山山脉西北端汇合处，紧靠大名鼎鼎的塞罕坝林场，滦河最大的支流小滦河发源于此。这里森林草地相间，在斜阳下显得如此的静谧和美丽。

Lying near the southeast edge of the Inner Mongolia Plateau, the Yudaokou Ranch of Weichang County stands exactly at the intersection of the eastern end of the Yinshan Mountain Range, the southern end of the Daxinganling Mountains and the northwestern end of the Yanshan Mountain Range. Not far from the famous Saihanba Forestry Plantation, it is the place where Xiaoluanhe River, the largest tributary of the Luan River, has its origin. The forests rangeland basked under the slanting sunlight appears very tranquil and beautiful.

1-2 典型草原生态系统

TYPICAL RANGELAND ECOSYSTEM

呼伦贝尔高原、锡林郭勒高原、乌兰察布高原、鄂尔多斯高原、黄土高原，其海拔在800~1500米，其上分布着温带地带性的典型草原生态系统。由于区域气候主要受纬度、海拔、降水的影响，温性草原生态系统有着不同的类型，在内蒙古高原以耐寒草本植物为主，在黄土高原、鄂尔多斯高原以喜暖草本植物为主。

典型草原是世界著名欧亚大草原的代表性植被，年平均降水量为250~350毫米。这里逐水草而居的传统游牧生活，已有数千年的历史，这里草原生态系统不被破坏的可持续状态主要取决于放牧强度。游牧这种方式是在生产力水平较低的背景下形成的，对于草地资源的合理利用、维持草原生态系统的平衡有合理的作用。但是，随着人口的增加、经济的发展，人们对生活水平与生存环境的要求提高，要长期保持这种游牧方式已经不可能了。目前，这里的生活方式基本上已经变成定居式的，因此必须建立合理的放牧制度，在充分利用草地资源的同时，既不使草地资源遭到破坏，又能获得优质的畜产量，以满足人们不断增长的需求，是亟需解决的重大问题。

这里是京津风沙源的主要策源地之一，担负着建设中国北方生态屏障的最重要任务，这个问题的解决已经提到了刻不容缓的国家生态战略地位上来了。

What is extensively distributed on the Hulunbeier Plateau, Xilingol Plateau, Ulanchab Plateau, Ordos Plateau, and Loess Plateau, where the altitudes range from 800 to 1,500 meters, is the typical rangeland ecosystem that falls within the temperate zone. Given that the regional climate is often subject to the influences of latitude, altitude and precipitation, the temperate rangeland ecosystem may come in a fairly large array of different types. Whereas the Inner Mongolian Plateau is normally covered by cold-tolerant herbaceous plants, the dominating vegetation species on the Loess Plateau and Ordos Plateau might turn out to be those that thrive in warm places.

Typical rangelands are the most representative of the world-famous Eurasian steppe, where the average annual precipitation stands around 250-350 millimeters. The traditional nomadic lifestyle of this region — under which herdsmen normally settle down in places of abundant water and thriving grasses — can be traced back to several-thousand years ago. The long term sustainability of the rangeland ecosystem that exists widely here depends mainly on whether the intensity of grazing is properly controlled so as to keep the rangelands from being

草肥羊壮•内蒙古锡林浩特

Luxuriant grasses and fat sheep — Xilinhot, Inner Mongolia

横跨欧亚大陆的典型草原是世界草原之首，它的气候、植物、动物、土壤等最能代表草原的生态特征，因此被称为典型草原。自欧洲多瑙河起，呈带状连续往东延伸，经东欧平原、西西伯利亚平原、哈萨克丘陵、蒙古高原，直达中国东北。这里曾经演绎了匈奴、突厥、鲜卑、契丹、女真及蒙古若干草原王国伴随着、交融着中华文明数千年兴衰的历史。

Typical rangelands that stretch across the Eurasian continent rank the first among the world's rangelands. Since the climate, plants, animals and soil found here are the most representative of ecological characteristics of the rangeland ecosystem, they are normally referred to as typical rangelands. Starting from the Danube River in Europe, this vast steppe extends all the way eastward in the shape of a massive belt, passing through the plains of Eastern Europe and Western Siberia, through the Kazakh hills, through the Mongolian Plateau until finally reaching northeastern China. This is also the place that has in the history nurtured numerous rangeland-based kingdoms like Xiongnu, Turkic, Xianbei, Khitan, Jurchen and Mongolia, all of which form integral parts of the splendid Chinese civilization that has flourished for thousands of years.

damaged. Nomadic lifestyle, which has its origin in early days when the overall level of productivity was low, is a reasonable choice and plays a fairly positive role in the rational utilization of grassland resources and maintaining the balance of the grassland ecosystem. However, with the increase in population and economic development, people's demand for better standard of living and living environment has also increased, making it no longer feasible to stick to this nomadic pattern of living. Since the prevalent style of living here now is sedentary, an urgent task we are faced with is to put in place a reasonable grazing system that could, while making sure that the grassland resources are kept free from undesirable damage, enable people to make the fullest use of them by sustaining an optimal livestock population, hence satisfying the ever-increasing demands of people.

This is also the place from which sandstorms that occasionally hit Beijing, Tianjin and surrounding areas have their origin, and therefore shoulders important responsibilities to build an ecological shelterbelt for northern China. Proper solution of this problem has now already been elevated to the position of a national ecological strategy that needs immediate implementation.

夏草场·内蒙古呼伦贝尔
Summer pasture —
Hulunbeier, Inner Mongolia

草原文化中最有历史、最延绵不断的社会行为就是游牧，这是草—畜—人相互促进、相互制约、动态平衡的核心所在。我们现今的草原人口压力大，经济活动频繁，很多地方的牧民已经定居。牧民在冬、夏草场中转场，或四季转场的现象已经不容易见到了。

The longest-lasting and most-enduring social act of rangeland-based culture is the nomadic husbandry, an essential way of living shaped by the mutually-dependent, mutually-complementing and dynamically-balanced relationship between grasses, livestock and human beings. Due to the tremendous pressures exerted by ever-increasing human population and frequent economic activities, the nomadic herdsmen inhabiting many rangelands have now settled down to sedentary lives. The grand and impressive event when herdsmen and their animals move between summer and winter pastures, or between other seasonal pastures has already become a view that we can seldom see now.

蘑菇圈·内蒙古包头

Mushroom rings —
Baotou, Inner Mongolia

在草原上发现蘑菇，是一件让人非常欣喜的事。除了很多蘑菇如黄蘑等可以食用之外，有经验的牧民会告诉你，这是好草场的标志，也是过牧草场开始恢复的标志。

It is an exciting and pleasant thing for people to come across mushroom rings on rangelands. This is not only because the rich diversity of mushrooms available in the rings are highly palatable, but also because they serve, as well-seasoned herdsmen will tell you, as the precursors of good-quality rangelands, signifying that an overgrazed pasture is starting to recover from degradation.

金莲花·河北沽源

Trollius chinensis —
Guyuan, Hebei

金莲花喜冷凉湿润环境，多生长在海拔1000~2200米的高山草甸或疏林地带。金莲花有很好的观赏价值和药用价值。

Fond of cool and wet climate, *Trollius chinensis* are often found at the 1000–2000m-high alpine meadows or sparse forest belts. This beautiful flower is known for its high ornamental as well as medicinal values.

牛粪堆上的鸟•内蒙古通辽
Bird on cattle dungs —
Tongliao, Inner Mongolia

草原上的牛、羊等吃草后排出的粪便，会被鸟、昆虫等利用，再慢慢被蚯蚓等低等动物以及微生物等分解，变成了滋润草原的肥料，有力促进草的生长，从而形成养分的良性循环。图中站在粪堆上的鸟为田鹨，在中国分布较广，主要栖息于开阔草地、河滩、林缘灌丛、林间空地等。

On rangelands, birds and insects would make pretty good use of dungs discharged from the consumption of grasses by cattle and sheep before these "wastes" are gradually decomposed by earthworms and microorganisms and turned into fertilizers that nourish the rangelands as well as grasses growing there. In this way, a welcoming virtuous cycle of nutrients comes into being. The bird you see standing on cattle dungs in the picture is *Anthus richardi*, a species widely found across China and inhabiting primarily such places as open grasslands, riverbanks, forest margin thickets and forest clearings.

▼ 草垛上起飞的雪鸮•内蒙古呼伦贝尔
A snowy owl in haystacks ready to take wings — Hulunbeier, Inner Mongolia

在北温带，雪鸮主要生活在北极和西伯利亚，呼伦贝尔草原已经是雪鸮生存的最南端了。近年来，由于生态环境的好转，在这里越冬的雪鸮数量越来越多了。图中草垛子是草场割草打捆准备运走的。

Snow owls (*Bubo scandiacus*) in northern temperate zone mainly live in the Arctic and Siberia, with Hulunbeier rangelands being the southernmost point of their habitats. Over recent years, due to the much improved the ecological environment, an ever-increasingly number of snowy owls have come here to stay for winter seasons. The haystacks shown in the picture are harvested and baled grasses waiting to be moved away.

▲ 肥美草场牛群壮•内蒙古锡林浩特
Lush pasture and stout cattle — Xilinhot, Inner Mongolia

典型草原的代表性植物为针茅。针茅为多年生密丛类禾草，营养成分、适口性和耐牧性均很高，是优良的饲用植物，特别是其营养生长期粗蛋白质含量较高，是牛、羊、马都喜好的食物。

The plant species most representative of typical rangelands is *Stipa capillata*, a perennial and dense-growing herbaceous plant that is highly nutrient-rich, grazing-tolerant and palatable, making it an excellent fodder for livestock. Because of its especially-high crude protein content during growing periods, it turns out to be a most favored fodder for cattle, sheep and horses.

草原雕起飞•
内蒙古呼伦贝尔
An eagle about to take wings — Hulunbeier, Inner Mongolia

草原雕的名字取自草原，属大型猛禽，栖息于开阔草地、荒漠和低山丘陵地带的荒原草地，以草地上生活的黄鼠、跳鼠、沙土鼠、野兔、旱獭、沙蜥、蛇和鸟类等小型脊椎动物为食，是草原食物链上的顶级物种。

Rangeland eagle (*Aquila nipalensis*), whose name is derived from the place they live on, belong to large raptors that typically inhabit open grasslands, deserts and low hills in the wilderness, where they feed on small vertebrates such as weasels, mice, gerbils, hares, otters, sand lizards, snakes, and birds that abound here. They rank among the top species among the grassland food chains.

黄羊群过路•
内蒙古新巴尔虎右旗
Migrating *Procapra gutturosa* — New Barkhor Right Banner, Inner Mongolia

黄羊（学名蒙原羚）分布于中国、蒙古国交界以及俄罗斯与中蒙交界的部分地区，喜好群栖、群移，一般在春季和秋季进行大规模迁移，移动的距离和范围都比较大。图中是在中蒙边界来回迁徙的黄羊群在经过汽车道路时的状况。从图中可以看到，草层是比较浅薄的，被汽车压坏的地方就是沙质土层。草原的退化、沙化就是草层退化或者遭到破坏的结果。

Procapra gutturosa is widely found in the border areas between China and Mongolia as well as in parts of places sitting on borders between Russia, China and Mongolia. They tend to live and migrate in groups, covering long distance and extensive scopes of land during the spring and fall seasons when they go on massive migrations. What is shown in the photo is the view when the herds that migrate regularly along the China-Mongolia border are crossing highways. It can be seen from the photo that the grass coverage is relatively shallow, revealing bare sandy soil in places where vehicles pass. Rangelands degradation and sandification would happen when the grasses are damaged or get thinner.

草原风云•
内蒙古呼伦贝尔
Unpredictable weather over rangelands — Hulunbeier, Inner Mongolia

在夏季广阔无垠的蒙古大草原上，强烈的对流天气使草原上空变幻莫测，一边阳光灿烂，一边浓云滚滚；一时狂风暴雨，一时微风和煦，阳光、乌云、大风、骤雨组合拳一阵乱舞。冬天的自然环境更加恶劣。因此，草原上牧民的生活条件非常艰辛，尤其是在游牧的状态下。

Due to the influences of strong airflows over the boundless Mongolian rangelands, the weather here in the summer may often turn out to be hardly predictable. Whereas this part of the rangelands is bathed in bright sunlight, another part may already be shrouded by thick dark clouds; whereas it may be ridden by gales and downpour at this moment, balmy breeze and warm sun would emerge in the very next few minutes. Bright sunshine, dark clouds, strong winds and torrential rains would often join hands to present you with something that you can hardly tell what sort of weather it is. The natural environment is even harsher in winter, so the living conditions of herders on rangelands are very difficult, especially in the nomadic state.

天光下的马群·
内蒙古巴彦淖尔
Herds of horses under the sky — Bayannur, Inner Mongolia

马曾经是内蒙古草原上的主要畜种之一，随着社会的进步、经济的发展，马的代步行走和运输功能逐渐被现代化的摩托车、汽车所替代，养马的投入产出比下降了。因而，马在“草原五畜”（绵羊、山羊、马、牛、骆驼）中所占的比例也在逐步下降。

Horses used to be the predominant type of livestock on the Inner Mongolian steppe. However, with the progress of society and economic development, the roles played by horses — as means of transportation and beast of burden — have gradually been replaced by modern motorcycles and automobiles, significantly cutting down the input-output ratio of horse-keeping. As a result, the proportion of horses among the “five prevalent domestic animals on rangelands (sheep, goats, horses, cattle and camels)” is also declining by and by.

1-3 沙化草原生态系统

SANDIFIED RANGELAND ECOSYSTEM

在我国半干旱半湿润的内蒙古高原草原地区，包括东北丘陵草原地区，有不少沙化草原分布，如在浑善达克沙地、科尔沁沙地、毛乌素沙地等。这里光、热、水条件都较好，但植被覆盖率较低，这些沙地的形成主要是因为人类过度利用甚至开垦造成的，这是近现代人类不合理开发使中国北方草原退化的不良结果。

这部分沙化草原生态系统并非是地带性的、原生的生态系统，而是草原生态系统退化、沙化的结果。无论是从历史上原生植被的角度看，从草原退化的现实给人类的警示看，还是从全球气候变暖、气候带北移的客观趋势来看，我们都有信心将这些退化的草原生态系统恢复成它原本的、稳定的地带性草原生态系统或次生草原生态系统，无论如何我们都不应该将这部分沙化草原的成因遗忘。如果我们遵循“师法自然”的生态修复方式，扼制住荒漠化进程，这些地方优越的自然地理气候条件完全可以将这片土地恢复成原本的典型草原生态系统或森林草地生态系统（也包括次生生态系统，下同）。我们近些年在这块地方采取的防沙治沙和草原生态修复实践的成果已经证明了这一点。

总之，切不可将这片可以恢复成原来草原生态系统的区域僵化地划到荒漠生态系统中去，这是该区一个方向性的问题，必须得到充分地重视并解决。

There are some sandified rangelands in China’s semi-arid, semi-humid Inner Mongolia Plateau Rangeland Region, even in the hilly north-eastern rangeland region, with such sandy lands as Hunshandak, Horqin, Mauwusu being a few of the examples. These places are actually endowed with fairly favorable light, heat and water conditions, but the vegetation coverage thereof is comparatively lower. The existence of these sandy lands is mainly caused by the abusive utilization and over reclamation of these lands by people. It is the irrational use and over-development of these rangelands by modern human beings that has caused the undesirable degradation of such places in northern China.

The sandified rangeland ecosystem that exists today is not the zonal one that has always been like this since the very beginning; rather, it is the result of degradation and sandification that previous rangeland ecosystem had suffered. Whether judging from the perspective of the primitive vegetation that existed here in history, from the alarming bells that nature has sounded for us human beings concerning rangeland degradation, or from the objective trend of global warming and the northward shift of climate zones, we are quite confident that we will be able to restore these degraded rangeland ecosystems or secondary grassland ecosystems back to their previous healthy and stable state. Whatever the case is, we should by no means forget the real causes of these sandified rangelands. So long as we make efforts to curb the process of desertification and adopt concepts and measures for ecological restoration that “fit in well with the natural mechanism”, the natural geographic and climatic conditions of these places will be sufficient to restore these lands to the typical rangeland ecosystem or the forest grassland ecosystem (including secondary eco-systems. The same principle applies in the following sections). This has been proved by our practices in sand prevention, sand control and rangeland ecological restoration in these places over the years.

To sum up, it will never do to rigidly consign these areas that might well be restored to their previous states to the desert ecosystems. This is a matter about what direction we will be following and therefore must be given its due share of attentions.

风向与沙向·内蒙古通辽
Directions of wind and sand — Tongliao, Inner Mongolia

科尔沁沙地这个地域风旱同季，风向固定，沙向就固定，风与沙的关系在这里演绎得如此地合拍、如此地淋漓尽致。这里属于半湿润半干旱地区，这种景观是原有植被遭到人类破坏后，土地沙化后的结果。

The seasons of wind and drought coincide in the region where the Horqin Sandy Land lies. Where the direction of prevalent winds is fixed, that of sands will also be fixed. The well-synchronized relationship between winds and sands finds its most telling, most vivid illustration here in the photo. This area belongs to the semi-humid semi-arid region, where the view can only be explained as the consequence that results when the original vegetation is so seriously damaged by human-induced activities as to lead to sandification of the land.

离北京最近的大沙源·
内蒙古苏尼特左旗

A major sandstorm source that is the closest to Beijing — Sunit Left Banner, Inner Mongolia

受东亚季风及环境变迁的影响，这里是距离北京最近的沙地——浑善达克沙地，在沙丘间发育着疏林、灌丛和草地，有固定或半固定沙丘，丘间多甸子，主要由浅黄色的沙组成，其景观在温带荒漠草原和森林草原之间波动变化，历史上出现过一系列沙地扩展与沙丘固定、沙地收缩与沙丘活化的推拉波动现象。它坚定地在告诉人们：这块土地“毁之则荒、护之则绿”。

Due to the influences of the East Asian monsoon and that caused by environmental changes, sparse forests, scrubs and grasslands are appearing amidst the sand dunes of the Hunsandak Sandy Land, the nearest among its type to Beijing. Both fixed and semi-fixed dunes can be found here, between which are patches of fields composed mostly of light-yellow sands. Landscapes here in history have alternated frequently between that of temperate desert rangelands and that of forest rangelands, leaving behind it a fluctuating trajectory characterized by concurring desert expansion and dune fixation, or contracted desert and loosened dunes. This makes a convincing case that reminds us the following binding fact: This is a land that “can turn barren when it is used abusively or flourish when properly taken care of”.

跟着妈妈走的小石鸡•内蒙古巴彦淖尔

A partridge hen and her chicks — Bayannur, Inner Mongolia

石鸡广泛分布于中国的三北（西北、华北、东北）地区，常栖息于低山丘陵地带的岩石坡和砂石坡上，主要以草本植物和灌木植物的嫩芽、叶、果、种子，以及苔藓、地衣和昆虫为食。图为石鸡母亲带着一群孩子正在山坡上觅食。

The rock partridge (*Alectoris chukor*) is widely found to inhabit the rocky and gravel slopes in hilly areas or low mountains in northwestern, northern and northeastern China. It mainly feeds on the shoots, leaves, fruits and seeds of herbaceous and shrubby plants, as well as on mosses, lichens and insects. Shown in the picture is a brood of partridge chicks foraging behind their mom on a hillside.

沙地花棒•内蒙古包头

Corethrodendron scoparium on sandy lands — Baotou, Inner Mongolia

细枝羊柴（别名花棒）为沙生、耐旱、喜光的半灌木种，主、侧根系均较发达。它适应生长于流沙环境，喜沙埋，抗风蚀，耐严寒酷暑，叶枝茂盛，萌蘖力强，防风固沙作用大。

Corethrodendron scoparium (literally also known as flower stick) is a sand-borne, drought-resistant and light-prone semi-shrubbery species that has fairly developed main and lateral root systems. Well-adapted to environment consisting of quick sand, it can flourish in sand- and wind-ridden places where the summer and winter could be extremely harsh. Its leafy and robust stem comes with very strong regenerating power, making it a highly valuable wind-breaking and sand-fixing plant.

草原明沙·内蒙古赛罕乌拉
Pseudo dunes over rangelands —
SaihanWula, Inner Mongolia

这里是内蒙古赛罕乌拉国家级自然保护区的王坟沟生态景观，一道道明沙丘是这片草地曾经发生过的土地严重沙化的忠实记录。自二十世纪末自然保护区成立后，采取了保护、封禁、生态修复等诸多措施，明沙丘停止了活化并逐渐固定了下来，沙丘逐渐被植被覆盖，沙丘底部水分好的地方还长出成片的乔木。这让我们看到了这类草原生态修复的美好前景。

Shown in the photo is a view taken at Wangfenggou scenery spot that is located within the SaihanWula National Nature Reserve in Inner Mongolia. The row upon row of pseudo dunes bear faithful witness to the fact that this stretch of grasslands had once been subjected to serious sandification. Following the establishment of the nature reserve at the end of the twentieth century, a series of measures have been taken to restore the ecology. As a result, the active dunes have gradually slowed down in movement and become fixed ones, upon which vegetation began to appear. In cases where the base part of dunes is endowed with favorable water conditions, strips of tall trees have emerged. This encouraging sign promises us a bright prospect for ecological restoration in such type of rangelands.

榆树沙地·内蒙古通辽
Elm-dotted sandy land —
Tongliao, Inner Mongolia

在科尔沁沙地，榆树和沙丘、草地构成了一类特殊的沙地景观，这里光、热、水条件都比较好，如果遵循“师法自然”的理念，停止伤害现有植被，采取人工促进、生态修复的方法，这些沙化土地是完全可以修复、回归到典型草原，甚至回归到森林草地状况的。

In the Horqin Sandy Land, elms, dunes and grasslands combines to present a very special type of sandy landscape. Endowed with fairly favorable light, heat and water conditions, this type of sandified place is highly possible to be restored to typical rangelands, even to forest grasslands, so long as proper ecological restorative measures are taken while we strictly follow the “go-after-nature” philosophy and stop damaging existing vegetation.

1-4 草甸生态系统

MEADOW ECOSYSTEM

草甸生态系统以多年生草本植物为主，它并不像草原生态系统那样连续分布，而是间断条块状分布。这是一种特殊的草地生态系统类型。

草甸生态系统的显著特征是有变动的地下水或地表土含水量高，土壤类型为沼泽草甸土、泥炭草甸土等，主要是水的增多，使地带性生态系统向典型草甸生态系统、沼泽草甸生态系统向沼泽湿地生态系统变化产生的（沼泽湿地生态系统已属于湿地范畴，参见“一滴水生态摄影丛书”之一的《多样性的中国湿地》，本书不再涉及）。

在一些地势低洼的地区，还分布着一些面积较大的盐碱土或盐化草甸土。这类土壤含盐量较高，地下水较浅，土壤经常比较湿润，或有短期的季节性积水，在这种条件下，只有比较耐盐的或盐生、泌盐的植物才能正常生长，形成盐生草甸类型的生态系统。

在中国，草甸生态系统广泛分布于不同的气候带，因其所处地的纬度、海拔、水分、植物种类的构成不同而不同。由于草甸生态系统为广布类型，本书所有章里都会涉及，以上介绍属共性的内容，后面就不再重复。

本区主要分布有典型草甸生态系统、沼泽草甸生态系统与盐生草甸生态系统三类。

The meadow ecosystem is dominated by perennial herbaceous plants. Instead of extending in continuous stretches like commonly-seen rangeland ecosystem, it tends more often to appear in fragmented stripes or patches, thus making it a very special type of grassland ecosystems.

A most notable features of meadow ecosystems is the presence of groundwater or surface water, making the soil thereof fairly high in water-content. Soils found here are normally of swampy meadow soil, peat meadow soil and so on. A key cause for their existence is the increased availability of water, which shifts zonal ecosystem to typical meadow ecosystem, and swampy meadow ecosystem to swampy wetland ecosystem (the last type already belongs to the category of wetlands, and therefore will not be covered in this book. Readers interested can refer to *Diverse Wetlands of China* that has previously been published under the “A Single Drop of Water Ecological Photography Album Series”) .

In some low-lying areas, vast stretches of saline soil or salinized meadow soils might also be found. This type of soil often comes with high salt content and shallow groundwater, hence making the soil constantly wet or prone to seasonal short-term waterlog. Only plants that are comparatively more salt-tolerant or by nature salt-borne can flourish under this kind of condition. Therefore, it belongs to the type of saline meadow ecosystem.

The meadow ecosystem is widely distributed in different climatic zones across China, differing somewhat from each other depending on the latitude, elevation, moisture, and plant composition of their specific locations. Since the meadow ecosystem is a type that exists extensively and will be covered in all chapters, the general introduction concerning its shared features mentioned above will not be repeated in later chapters.

This region is home to three types of major ecosystems: the typical meadow ecosystem, the marshy meadow ecosystem and the saline meadow ecosystem.

草原中的沼泽地•内蒙古阿巴嘎旗
Swamps in rangelands — Abaga Banner, Inner Mongolia

大沙梁洼地里汇集了来自四面八方的水分，在最低洼处形成了湖泊，随着坡地逐渐上升，土壤含水量逐渐减少，湖泊湿地生态系统逐渐向沼泽湿地生态系统变化，并逐渐向沼泽草甸生态系统、典型草甸生态系统方面转变。

Water from all directions gathers at the lower lands of Dashaliang to form a lake. As the slopes rise in elevation, water content in soils decreases increasingly, causing the lake wetland ecosystem to change progressively towards the marshy wetland ecosystem, the marshy meadow ecosystem, and finally to typical meadow ecosystem.

草地河流与沼泽·
内蒙古兴安盟
River and swamp in grasslands — Xingan League, Inner Mongolia

在平缓的大草原上，由于终年冻土层的存在，河流下切并不明显，因此形成这样曲曲弯弯、千迴百转的河床，河床还常常改道，甩出一段又一段的牛轭湖。这里的草地饱含水分，形成了大片的典型草甸和沼泽草甸。

Due to the existence of permafrost layers on the flat rangeland, the river that meanders across it does not cut obviously into the ground, only leaving behind it a shallow riverbed that turns, twists, winds and bends. Moreover, the riverbed sometimes diverts abruptly to bring into existence one after another oxbow-shaped lakes. This water-rich grassland here gives birth to vast stretches of typical meadows and marshy meadows.

秋色羊群·
内蒙古根河
Sheep under the autumn sky — Genhe, Inner Mongolia

秋天来了，染黄了桦树和草原，静静流过的河水变得更加平缓，映照出天空的湛蓝。草甸之中，羊群正抓紧上膘，准备熬过中国北方草原长长的冬天。

As the arrival of autumn turns the rangeland and birches into a golden sea of bright colors, the river that meanders quietly across it also becomes ever more tranquil and slow-paced, adding further to the transparent blueness of the sky higher up. Countless sheep are busy feeding themselves so as to be prepared for the long and harsh winter of the northernly rangelands.

碱蓬草地•
宁夏银川
A grassland of *Suaeda glauca* — Yinchuan, Ningxia

碱蓬是一年生的盐生草本植物，俗称“盐蒿”，成熟时植株火红。它粗壮的根系能够在含盐量比较高的滩涂土壤里扎根。盐地碱蓬的生长与它所扎根的滩涂土壤的含盐量密切相关。这是大片盐生草甸生态系统的景观。

Suaeda glauca, an annual herbaceous plant that grows in saline condition and hence known commonly as "*Artemisia salina*", would turn brightly red when it reaches its maturity. Its sturdy root system is capable of settling down and flourishing in mudflat soils that is of relatively high salt content. The growth of the salt meadow is closely related to the salt content of the mudflat soil in which it is rooted. Shown in the photo is the view you can expect to see on a vast stretch of saline meadow ecosystem.

塔头披纱•
山西大同
"Pagoda-head" behind veils — Datong, Shanxi

为适应地表过湿或积水环境的变化，薹草生长比较密集，其根系迭生产生了地表起伏，形成了草丘微地貌（又称"塔头"），看起来像一个个争先恐后发酵的馒头。薹草草丘作为一种独特的景观，广泛分布于北方的泥炭沼泽草甸中。在大雪覆盖下，塔头地就像披上了一层白纱，已经没有了往日"争抢"的劲头，都在静静地等待着春天的到来。

To better adapt to excessively-wet or waterlogged environment, *Carex* species tend to grow in dense groves, where the tightly-clustered root systems of them would often cause the ground to elevate into an undulation that resembles the head of a pagoda (hence commonly known as "pagoda-head"), or looks like one after another of fully fermented steamed buns. This unique landscape made of tightly-clustered *Carex* roots is widely seen in marshy peat meadows in northern China. When snow falls in winter, it would look as if the "pagoda-heads" had put on some eye-catching white veils that transform them into gentle and quiet brides who are looking eagerly forward to the arrival of spring.

1-5 可持续发展

SUSTAINABLE DEVELOPMENT

内蒙古高原的草原生态系统受人类活动影响的时间长、强度大，人类的干扰在本区不同区域表现的形式也有所不同。长期以来，本区是中国最重要的畜牧业基地，不仅内部循环，还需要向外供给国家大量肉、奶、皮、毛等畜产品。因此，该区普遍存在且不容易解决的问题是过度放牧。而该区东部和南部水热条件较好的地方，又被当作潜力巨大的宜农宜林地来开垦，致使优良草场的面积在不断地缩小、被侵蚀。开矿和对草原植被的乱采滥挖（尤其是西部）对于草原生态系统的危害也是致命的。

本区为中国动物地理区划中古北界蒙新区的东部，即蒙古高原、鄂尔多斯高原区。这里分布着黄羊、驯鹿等食草动物，狐狸、草原雕、大鸨等野生动物也在这里生存、繁衍，这里有众多的迁徙鸟和留鸟，如雪鸮、鹤类等。草原植被主要由针茅、羊草、金莲花等禾本科、菊科、豆科植物组成。

本区生态系统服务和生态功能的核心是土壤、植被及生产力的可持续稳定、草畜产品的提供、防风固沙和生物多样性保护。在实施措施上，应坚持生态优先、适度利用的原则，针对不同草原生产力，分类、分区实现不同自然条件下的以草定畜政策，严格控制载畜量，遏制超载放牧、草地退化的势头。要在不同地区采取相适宜的禁牧、休牧、减牧等管理措施，划出禁牧区和禁牧期；要制定好轮牧制度，合理轮牧、划区轮牧，要变不合理的网围栏分割的“小草原”为便于轮牧的“大草原”；要从根本上协调解决好草—畜—人的矛盾、自然保护与生产生活的矛盾；要从中国北方生态屏障的防沙治沙、自然保护和生态修复的战略高度来充分认识和开展工作。

要加大生态修复的力度，对沙化、退化草场尤其要“师法自然”，采用乡土草种补播、土壤生物修复等措施，加速风蚀、水蚀、水土流失、盐渍化草地的生态修复，用人工促进天然更新的办法恢复草场生态，应将一部分生态空间和承载力还给野生动植物；严格限制矿产资源开发，禁止开垦草原；创新草牧业发展方式，适度发展草原生态旅游业，探索草原生态保护与草牧业高质量发展的新模式。

经过长期的自然环境选择与人类培育，“草原五畜”（牛、马、绵羊、山羊、骆驼）是本区最重要的畜牧品种，只是从东到西不同的草原在五畜比例上略有差别，不同的畜偏食不同的草，这样有利于草原的可持续利用。本区还生存有许多具有特殊价值的珍稀动植物种，如黄羊、草原雕、大鸨等，植物多样性异常丰富，因此这里是最为难得的典型草原生物基因库。

此外，在森林草原交错区、草原荒漠过渡区，在自然保护和生态修复方面，一定要遵循“师法自然”的原则，宜林则林、宜灌则灌、宜草则草、宜荒则荒，切不可为一时的所谓“中心任务”而做出违背自然规律的事来。

下一步要努力建设好以呼伦贝尔、大兴安岭、燕山—塞罕坝、太行山、大青山、贺兰山等草原类型和涉草类型国家公园为主体的自然保护地体系，完善草原类自然保护地空间布局和建设；加强中—蒙—俄跨国野生动物生态廊道建设，构建好以自然保护地为重要支撑的祖国北方生态安全屏障体系。

The rangeland ecosystem in Inner Mongolian Plateau Region has long been subjected to intensive impacts of human-induced activities. However, the interference of human beings also varies notably in the way it presents itself in different places. For a fairly

草原累了·内蒙古呼伦贝尔
Fatigue-afflicted rangelands — Hulunbeier, Inner Mongolia

这里是有名的呼伦贝尔七仙湖草原。传说中，这里美丽得让天上的七位仙女都被吸引住，产生要下凡冲动的地方而得名。在航拍的角度，我们可以看到这片草原上羊群密度很大，草原已经呈现出退化、沙化、盐渍化的迹象。草原累了，真的需要修生养息了。

Shown in the photo is the highly reputed Qixianhu (Seven Fairy Lake) rangeland in Hulunbeier. Legend goes that this site used to be such a beautiful place that the seven fairy sisters in the heaven were so much intrigued by it that they were even willing to give up their immortal status in exchange for the privilege to live here. Viewed from the air, it can be seen that this rangeland is intensively grazed by sheep herds, revealing signs of degradation, sandification and salinization. It seems to be really worn out and urgently needs recuperation.

long period of time, this region has been regarded as the most crucial base of China's husbandry industry, which was expected not only to meet its internal needs, but also to satisfy the external demands of the country in the supply of meat, dairy, hide and woolen products. For this reason, a prevalent yet not-easily-solvable problem that this region is commonly faced with is over-grazing. To make it worse, the eastern and southern parts of this region, due to the favorable water and heat conditions they are endowed with, have often been taken as high-potential sites and hence reclaimed for agricultural and forestry development, leading in consequence to ever shrinking and declining quality rangelands. Mining and unregulated excavation of rangeland vegetation (particularly in the western part) also pose fatal threats to the health of the rangeland ecosystem.

In terms of China's zoogeographical zoning, this region falls within the eastern part of the Inner Mongolia-Xinjiang district of the palearctic realm, which roughly coincides with the Mongolian Plateau and Ordos Plateau. Widely distributed here are herbivores like antelopes and reindeer, as well as large predatory animals like foxes, rangeland eagles, upland buzzards and others. This region is also home to countless migratory and resident birds, snowy owls and cranes being a few examples. Vegetation here mainly consists of plants belonging to the Gramineae, Compositae and Leguminosae families, *with Stipa capillata Leymus chinensis*, and *Trollius chinensis* being a few examples.

The core ecosystem services and ecological functions provided by this region include: sustainable and stable soil vegetation and productivity; provision of grass and livestock products; windbreaks, sand fixation, and biodiversity conservation. In terms of implementation, the ecological-priority, moderate-utilization principles should be strictly adhered to. Zone-specific, type-specific policies that take the productivity and carrying capacity of each type of rangelands into considerations must be worked out so as to effectively curb the trend of over-grazing and grassland degradation. In line with the specific conditions of each place, purpose-designed management schemes — grazing ban, moratorium and mitigation for instance — should be adopted to spell out clearly-stated geographical scope and time span of grazing ban; A sound rotational grazing system should be formulated to make sure this measure is applied to its best effects as well as to the most suitable sites; Measures must be taken to transform the irrationally-fenced-off "grassland patches" into "vast rangelands" where rotational grazing can be applied to its best effects; Well-coordinated actions must be taken to address the conflicting needs of grasses, livestock and people, and that between nature conservation and productive activities at their roots; We must fully understand and work from the strategic height of sand prevention and sand control, nature protection and ecological restoration of the ecological barrier in northern China.

Efforts must be strengthened for ecological restoration. In dealing with sandified and degraded pasturelands, the "go-after-nature" philosophy must be strictly followed through adopting native grass species, using bio-based soil restoration methods and *etc*. in the process. Efforts must be taken to speed up the ecological restoration of wind-eroded, water-eroded, soil-eroded and salinized grasslands through adopting human-enhanced natural regeneration initiatives. A certain proportion of the available eco-spaces and corresponding carrying capacity should be given back to wild animals and plants. Development of mineral resources on rangelands must be put under

strict regulation and duly restricted. Innovative manners must be taken to safeguard the long-term and sustained development of the rangeland and husbandry industry, for instance by encouraging moderate and well-planned grassland eco-tourism, so as to explore a new developmental model that meets the needs of both rangeland ecological conservation and high-quality husbandry industry.

As a result of natural selection and human cultivation, the "five grassland livestock (namely cattle, horses, sheep, goats and camels)" have been proven to be the key livestock species that best suit this region, only that the ratio between them varies slightly depending on the specific type of rangelands that are found from the east to the west. This makes perfect sense and is conducive to the long-term sustainability of rangelands. In addition to the rich diversity of floral species it boasts, this region is also the home of many rare and precious wild fauna species like antelopes, rangeland eagles, upland buzzards and *etc*., which explains why it is often credited as a hard-to-win biological gene pool of typical rangeland.

In addition, in places where forests and rangelands border on each other, or where rangelands gradually give way to deserts, any nature conservation or ecological restoration attempt must observe strictly to the "go-after-nature" philosophy. Rather than, for the sake of responding to the so-called "central tasks" of the time, rush blindly to something that obviously runs against the inherent laws of nature, we must use our own discretion to foster whatever ecosystem that best suits the local geographical and climatic conditions of the place at issue, be it that of forests, shrubberies, rangelands or deserts.

The next step we need to take is to put in place the natural protected area system that centers around the rangeland-featuring national park that covers the vast areas of Hulunbeier, Daxinganling, Yanshan-Saihanba, Taihangshan, Daqingshan and Helanshan, thus optimizing the spatial layout of rangeland-featuring natural protected areas. Efforts must also be strengthened to build a wildlife ecological corridor along the Sino–Mongolian–Russian borders, so as to put in place a sound system of ecological security shelterbelt in northern China.

长城边，古道旁，荒草凄凄黄•宁夏盐池

A grass-covered ancient trail by the Great Wall — Yanchi, Ningxia

中国农耕和游牧的边界大体在400毫米等降水量线，在它的两侧，历史上形成了截然不同的生态环境和文明体系。由于自然地理及历史文化的演变，这条线还与我国的长城走向大致重合。在目前全球变暖的大背景下，我国西北、东北、华北地区有气温升高、降水增加、呈暖湿化的趋势。"人努力天帮忙"，我们一定要顺势而为，生态修复要遵循"师法自然"的理念，努力建设好我国牢固的北方生态屏障。

The borderline between the agricultural and nomadic husbandry cultures in China runs roughly parallel to the 400-millimetre precipitation line, on either side of which distinct ecological environments and civilization systems have emerged over the history. Due to the evolution of natural geography and cultural history, this line also roughly coincides with the course of the Great Wall. Against the current background of global warming, the northwestern, northeastern and northern parts of China all seem to be on a trajectory of becoming warmer and wetter, as shown by the notably increasing temperatures and increased precipitation. Blessed with this favorable opportunity afforded to us by both nature and human efforts, we must exert our utmost to build up a solid and reliable ecological shelter in the north, bearing in mind the overriding "go-after-nature" philosophy at each step during the process.

CHAPTER 2 第二章

西北山地盆地草原区

The Mountainous and Basin Northwestern Rangeland Region

天山深秋鸟瞰•新疆阜康
A bird's eye view of Tianshan Mountains in late autumn — Fukang, Xinjiang

深秋的天山深处，河道露出了大片的河滩，金黄的草场成为了此时的大地底色，常绿的乔木和褐色的灌木点缀其中，它们在告诉人们，天山是干旱新疆腹地的“水塔”，是生物多样性的保存地，是当地老百姓生产生活的保障，是新疆中部最强大的生态屏障。

It's late autumn in the depth of the Tianshan Mountains, the river has receded to reveal huge patches of the riverbed. Vast stretches of pasturelands have by now turned brightly golden in color, dotted here and there with a few randomly scattered evergreen trees and brown shrubs. This view reminds us that the Tianshan Mountains — the "water tower" for the arid heartlands of Xinjiang as well as the most reliable ecological shelter of central Xinjiang — is both a biodiversity sanctuary and a safeguard of local folk's livelihood.

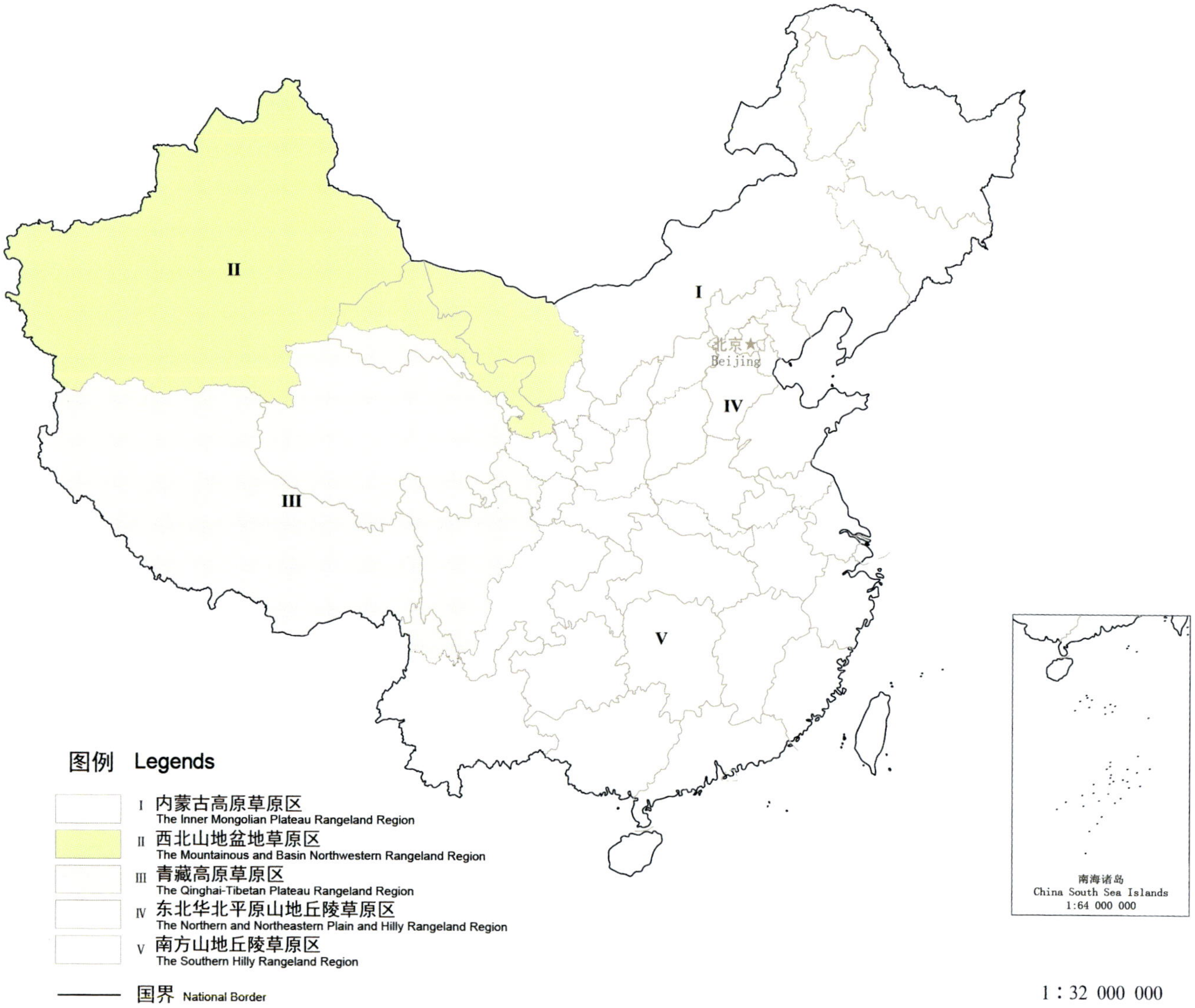
II
I
北京
Beijing
IV
III
V
南海诸岛
China South Sea Islands
1:64 000 000
图例 Legends
I 内蒙古高原草原区
The Inner Mongolian Plateau Rangeland Region
II 西北山地盆地草原区
The Mountainous and Basin Northwestern Rangeland Region
III 青藏高原草原区
The Qinghai-Tibetan Plateau Rangeland Region
IV 东北华北平原山地丘陵草原区
The Northern and Northeastern Plain and Hilly Rangeland Region
V 南方山地丘陵草原区
The Southern Hilly Rangeland Region
国界 National Border
1∶32 000 000

本区为东起狼山、贺兰山，南为青藏高原北缘的中国西北大部分国土，属温带干旱、极干旱的大陆性气候，包括贺兰山西坡、祁连山北坡、阿拉善高原、河西走廊及新疆的绝大部分地区，如天山山地和阿勒泰山地、塔里木盆地、准噶尔盆地和帕米尔高原等。

在这片广阔的温性草原生态系统分布区域内，受温度、水份因素的影响，分布着森林草地生态系统、干旱荒漠草原生态系统、极干旱荒漠草原生态系统和高寒草甸生态系统。有阿勒泰、天山、昆仑山等几座突起的巨大山脉，在这些山脉上，随着海拔高度的升起，出现草原生态系统类型的变化，如阿尔泰山、天山等区域，基部为干旱荒漠草原生态系统，向上为森林草地生态系统，山顶为高寒草甸生态系统。

此外，本区还分布草甸生态系统的若干类型。

This region, which covers most parts of China's territory in the north-west that is bordered by the Langshan and Helanshan Mountains on the east and the northern edge of the Qinghai-Tibetan Plateau on the south, belongs to the temperate continental climate zone that is mostly either arid or very arid. Specifically speaking, it mainly consists of the western slopes of the Helan Mountains, the northern slopes of the Qilian Mountains, the Alashan Plateau, the Hexi Corridor, and most parts of Xinjiang — the Tian Shan Mountains, the Altay Mountains, the Tarim Basin, the Junggar Basin and *etc.* for instance.

Due to the influence of highly varied temperature and water availability, a wide array of other ecosystems — the forest grasslands ecosystem, dry desert rangeland ecosystem, extremely dry desert rangeland ecosystem and alpine meadow ecosystem — can also be seen in this vast region that is otherwise predominated by temperate rangeland ecosystems. As the towering mountains like the Altay, the Tianshan and the Kunlun rise in elevation, the rangeland ecosystems would also change vertically. Take the Altay and the Tianshan mountains for example, the bottom parts are of the dry desert rangeland ecosystem, the top parts are of the alpine meadow ecosystem, whereas the parts in between are of the forest grasslands ecosystem.

Besides, several types that fall within the category of meadow ecosystem can also be seen in this region.

2-1 森林草地生态系统

FOREST GRASSLAND ECOSYSTEM

在本区，森林草地生态系统是森林生态系统向干旱荒漠草原生态系统过渡的一种生态系统类型，尽管这里已经是北冰洋、印度洋、太平洋暖湿气流影响的末端，但这里是我国西北最重要的牧场，森林和草地呈块状交替分布，镶嵌在山地的不同侧面和上下部位，大片草地主要分布于山坡的阳面与丘顶。

在西北地区，森林草地生态系统是所有生态系统中环境最良好的区域，也是草原生态系统中最湿润、生态条件最好的一种类型，主要分布于阿勒泰山、天山、祁连山的山体部分。这里降水量较高，土壤较肥沃，植物种类组成相对丰富，是我国温带干旱半干旱区里难得的草原生态系统，尽管林牧交错，但这里一直以来都是很好的宜牧区。在这种状况下，尤其要维护好该生态系统的平衡，放牧要控制在适当的密度上，更要严禁工程施工和过度放牧对森林自然更新造成的破坏，这对于保护生态环境、保障牧区人民生活、维护边疆稳定都有着极为重要的意义。

The forest grassland ecosystem that exists in this region is a type of transitional ecosystem that marks the progressive change from the forest ecosystem to the dry desert rangeland ecosystem. Although this region is situated already at the endpoint of places liable to the influences of the warm and humid air currents originated from the Arctic, the Indian and the Pacific Oceans, it still makes up the most important pasturelands in northwestern China, where forests and grasslands occur in alternate belts that are embedded at the various angles and elevations of the mountains. Large tracts of grasslands are typically found to exist at sunny mountain slopes or at the tops of hills.

Among all ecosystems found in northwestern China, the forest grasslands ecosystem normally occurs in places that are endowed with the most favorable conditions, and therefore also constitutes a type of rangeland ecosystem that is the wettest and has the best ecological conditions. These ecosystems are mainly distributed in the middle parts of the Altay, the Tianshan and the Qilian Mountains that come with high precipitation, fertile soil and rich plant diversity. They represent the rangeland ecosystem that is not readily seeable in the arid and semi-arid temperate zone of China. In spite that forestry and husbandry have both been key composing parts of local economy, this place has from the beginning been a very favorable site for developing animal husbandry. Under such a circumstance, attentions are especially needed to keep the balance of the ecosystem from being broken through controlling grazing activities at proper levels. More importantly, engineering projects and overgrazing must be strictly banned to avoid potential damages to the natural renewal of forests. This shall be of great significance to ecological environment conservation, to the livelihood of local pastoralists, and to the prosperity and stability of border areas.

天山森林草地雪•新疆新源
Snow-covered forest pasturelands at Tianshan mountains — Xinyuan, Xinjiang

横亘新疆全境的天山，长达1760千米，是准噶尔盆地和塔里木盆地的天然地理分界线。在年降水量超过400毫米的天山北坡中山带，有茂盛的森林。年降水量在400毫米以下的低山带，上部为良好的干旱草原（如伊犁地区的巩乃斯草原），下部为干旱荒漠草原。在过渡带有大量的森林草地镶嵌其中。

The 1,760-kilometer-long Tianshan Mountains, which stretches east-westwards across Xinjiang, is the natural dividing line between the Junggar Basin and the Tarim Basin. Whereas lush forests exist at the middle section of the northern slopes where the annual precipitation exceeds 400 millimeters, the belt of lower mountains where the annual precipitation is below 400 mm tends to be covered at the upper parts by flourishing good steppes (e.g., the Gongnais steppe in the Yili region) and by dry desert rangelands at the lower parts. Numerous forest grasslands are tucked in between at the transitional zones.

▲ 最后的净土•新疆博乐
The last piece of pure land —
Bole, Xinjiang

夏尔西里由于原来是军事争议区，没有人为干扰，生态环境基本保持原始状态，被称为新疆的"最后的一片净土"。这里高山带的高山草甸植被发育良好，中低山带阴坡为以天山云杉为主的森林草地带，阳坡为灌草型草原带。

Sharsiri, formerly a disputed military zone that has been kept from human interferences, boasts almost intact natural ecological environment and is therefore often reputed as the "last piece of pure land" in Xinjiang. Alpine meadows are especially well developed here at high mountains, whereas the two sides of medium- and low-lying mountain slopes are respectively enshrouded by belts of spruce-dominated forest grasslands and shrubbery rangelands.

▼ 夏尔西里的马•新疆博乐
A horse of Sharsiri —
Bole, Xinjiang

去往夏尔西里的途中，常常会偶遇特立独行的马，这种马肯定不是野马。那到底是人们散养的马还是丢失的马，尚不清楚。但是，这里有宽广肥沃的草地，在这里，马的自由生存一点问题都没有。

On their way to Sharsiri, people would often come across some solitarily roaming horses that are definitely not wild ones. Although we cannot say for sure whether they are horses set out by herdsmen to graze freely on the pastureland or ones that have gone astray by chance, what we do know is that, with the presence of such a vast and fertile pastureland, these free-ranging creatures will have no problem whatsoever in survival.

森林草场的精灵·新疆伊宁
A lovely creature of the forest pasturelands — Yining, Xinjiang

森林草场、林缘边，是新疆马鹿生活的最佳场地，它们在生长旺盛的草地里采食，一但有响动或者异常，撩开四腿就往森林里跑。图为雄性马鹿，其犄角相当粗大。

Forest grasslands and places at the edge of forests are the most favorable habitats for Tien Shan wapiti. These lovely and vigilant creatures will forage carefree over the verdant grasslands, but any slight disturbance or abnormality would send them on flee to the depth of forests. Shown in the picture is a male Tien Shan wapiti whose antler is especially large and robust.

雪地鹿群•
甘肃张掖
Deer roaming in the snow — Zhangye, Gansu

图中为在祁连山森林草地中生活的白臀鹿群，带头的雄性白臀鹿断了一只角。鹿的野外生存条件相当恶劣，打架斗殴、抵御天敌、钻林子碰岩石，均有可能被断角。

Shown in the picture is a herd of McNeill's Deer that inhabit the forest meadows of the Qilian Mountains. A part of the antler over the head of the leading male is missing. As the living conditions of deer in the wild are normally pretty harsh, the missing part is presumably lost while he was engaged in fights, trying to escape from predators, or was accidently knocked off by trees or rocks.

待转场的牛群•
新疆哈巴河
Cattle herds about to move — Habahe, Xinjiang

秋色已渐渐褪去，阿勒泰山的冬季即将来临，哈萨克牧民就要离开山上的夏草场，转场到山下的冬草场继续放牧。图为牧民正赶着牛群，准备启程。

As autumn draws to its end, winter will soon set in to the Altay Mountains. It is time for the Kazakh herdsmen to leave the pastureland on mountains and move to the low-lying winter pastureland. Shown in the picture are herdsmen busy ringing in their cattle and getting ready for the long journey.

2-2 干旱荒漠草原生态系统

DRY DESERT RANGELAND ECOSYSTEM

在中国的温带干旱半干旱区，干旱荒漠草原生态系统主要分布在内蒙古西部、甘肃中部和西部、新疆等地的狭长地带，呈东北—西南向分布，还分布在除了山地、沙漠、戈壁以外的其他地区。这里具有强烈的大陆性气候特点，日照充足，全年多风，蒸发强烈，降水量大都低于250毫米。

从东到西，随着干旱的加剧，典型草原生态系统逐渐过渡到干旱荒漠草原生态系统，此系统多分布于本区的平缓地区和山脉基部（山脉上部多为森林草地生态系统）。其植被类型以多年生丛生旱生或强旱生小禾草为主，并有一定数量的小半灌木与灌木组成的草地植被，其丰富度、草本植物高度、覆盖度及生物生产力明显低于典型草原。这里“五畜”之中的山羊、骆驼的比例较高。

在严酷的气候与土壤条件下形成的荒漠草原植被，呈现出低矮、稀疏、季相单调、生产力低的特点，旱生草本和灌木虽然稀疏但植株高大，保存有许多特殊植被成分。由于风蚀作用剧烈、太阳辐射强、地表干旱，这里土壤和细粒砂石留存时间短，地表多呈现粗砂和小砾石间有稀疏植被的景观。

In the temperate arid and semi-arid zones of China, the dry desert rangland ecosystem is mainly distributed in northeast-southwest direction in the narrow strip of lands in western Inner Mongolia, central and western Gansu and Xinjiang. Besides, it is also found in areas other than mountains, deserts and the Gobi. They show very obvious characteristics of continental climate — abundant sunshine, year-round winds, high evaporation, and less than 250mm precipitation.

The climate gets increasingly drier as we move from east to west, with typical rangeland ecosystem gradually giving way to that of dry desert rangeland ecosystem. This type of ecosystem in the region is mostly found in gentle slopes and the base parts of mountain ranges (the upper parts usually come with the forest grassland ecosystem). Vegetation here mainly consists of clusters of drought-tolerant perennial herbaceous grasses, miniature shrubs and semi-shrubs. Compared with typical rangelands, this type falls obviously short in the diversity, height, coverage and bio-productivity of plants. Among the “five key livestock”, the proportions of large-sized ones — goats and camels — are comparatively higher.

The vegetation of desert rangelands, in light of the harsh climatic and soil conditions under which it exists, tends to have the following defining features: low-growing, sparsely distributed, homogeneous in seasonal views, and relatively poor productivity. Nevertheless, in spite of their relatively scarce presence, the drought-tolerant herbs and shrubs tend to be tall in size and are in possession of many special botanical elements. As a consequence of intense wind erosion, strong solar radiation and arid surfaces, soils and fine-grained sands can rarely be retained for long here. Instead, the more frequently-seen view is that consisted of coarse sands and small gravels, over which sparse vegetation can occasionally be seen.

芨芨草草原•新疆哈密

Rangeland strewn with *Neotrinia splendens* — Hami, Xinjiang

芨芨草的生命力极强，耐干旱、耐盐碱，在其他植物不能生存的瘠薄土地上，仍可以茁壮地生长。春季，芨芨草的茎叶是牛、羊的最爱；夏季，芨芨草的茎部变得粗老，成为骆驼的美味；冬季，白雪将矮草覆盖，高大的芨芨草更成为饲草的主力；来年春季返青后，芨芨草生长的速度和生物量增长也很快。

Neotrinia splendens, a highly drought-and salinity-tolerant species, possesses very strong vitality and can thrive on barren lands where no other plants can survive. In spring, the tender stems and leaves of it are the favorite food of cows and sheep; As the stems turn coarse and thick, they become tasty delicacies for camels; As winter sets in and all other smaller grasses get covered beneath heavy snows, the tall *Neotrinia splendens* becomes the main source of fodder for all animals; When spring returns the next year, this plant will once again surprises people with the speed it grows and the fast increment in its biomass.

肉苁蓉•内蒙古阿拉善左旗
Cistanche deserticola —
Alashan Left Banner, Inner Mongolia

梭梭是一种多年生的小乔木，普遍生长于我国西北荒漠草原中。梭梭根部寄生有传统的珍稀名贵中药材——肉苁蓉，肉苁蓉具有较高的医用价值，被誉为“沙漠人参”。

Haloxylon ammodendron is a small perennial tree that is widely found in the desert grasslands in northwestern China. Parasitized at the root of this tree are a highly rare and valuable traditional Chinese herbal medicine — *Cistanche deserticola*, which is sometimes credited as “desert ginseng”.

阿尔泰盘羊•新疆富蕴

Argali —
Fuyun, Xinjiang

阿尔泰盘羊又称大角羊，为世界上体形最大的绵羊，习惯生活在在700~5200米山地草原和高山、亚高山高寒草甸草原。阿尔泰盘羊为群居动物，夏季常活动于雪线的下缘，冬季栖息环境积雪深厚时，它们则从高处迁至低山谷地生活，有季节性的垂直迁徙习性。多以禾本科、莎草科和葱属植物为食，也取食一些灌木的嫩枝叶。在交配期间，雄性阿尔泰盘羊争偶激烈，巨角相撞响声巨大。

Argalis(*Ovis ammon*), also known as the huge-horn sheep, are the largest sheep in the world and typically inhabit hilly rangelands or alpine and sub-alpine meadow rangelands that range between 700m and 5200m in altitudes. This gregarious species roams close to the snowlines in summer and moves downwards to low hills or valleys in winter when deep snows accumulate in their normal inhabiting niches, demonstrating vertical migratory patterns as seasons change. They primarily feed on Gramineae, Cyperaceae and *Allium* plants, but occasionally also forage tender shrubbery twigs. Fierce fights may often break out between male argalis in mating seasons, with the loud banging of their horns clearly hearable from far away.

旭阳草场•新疆阿勒泰

Sun-bathed pastureland in early morning —
Altay, Xinjiang

主要受北冰洋气候影响的阿勒泰山，这片新疆最北的山脉南坡比北坡干旱，山脉中上部是以西伯利亚落叶松、雪岭云杉等为主的森林草地，山脉下部是干旱的荒漠草原。哈萨克牧民在这里有着千年的游牧历史。

Subjected mainly to the influences of the Arctic Ocean climate, the lofty Altai Mountain that stands toweringly at the northmost part of Xinjiang is drier on its south-facing slopes than on its northern slopes. Whereas the middle and upper parts belong to forest grasslands dominated by Siberian larches and spruces, the arid lower parts consist mainly of desert rangelands. The Kazakh herdsmen have led a nomadic husbandry-based life here for thousands of years.

野驴群·新疆富蕴

Wild Asses —
Fuyun, Xinjiang

北疆中驰骋的野驴是蒙古野驴，和青藏高原的藏野驴相比，其毛色要浅黄些。蒙古野驴喜欢雄驴、雌驴和幼驴终年一起过游荡的生活，以禾本科、莎草科和百合科草类为食，极耐干旱，可以数日不饮水。蒙古野驴在干旱的环境中会找到合适的地方，用蹄刨坑挖出水来饮用，冬季主要靠吃积雪解渴。

Roaming freely over northern Xinjiang are Mongolian wild asses(*Equus hemionus*), whose fur is, in comparison with their counterparts that inhabit the Qinghai-Tibetan Plateau, of lighter yellow color. Mongolian wild asses tend to stay year-around in groups, with their spouse and younger ones hanging together all the time. They feed mainly on grasses of the Gramineae, Cyperaceae and Liliaceae families and is extremely resistant to drought, capable of going for days without drinking water. Mongolian wild asses have the abilities to find places in dry environment where they could dig water with their hooves. In winter seasons, they may also resort to snow to quench their thirst.

雪地三兄弟·新疆奇台

Buddies in the snow —
Qitai, Xinjiang

在卡拉麦里的茫茫雪地上，三只鹅喉羚小公羊出来寻找食物，这时干枯的小灌木枝条也是充饥的上品。

Three young male *Gazella subgutturosa* are foraging the boundless and snow-covered Kalamari. At this time of the year when food is scarce, even the dry twigs of small shrubs can become tasty foods to fill their stomachs with.

起飞·新疆吉木萨尔
Take wings —
Jimushar, Xinjiang

这片荒漠草原中，猛禽种类不少，如大鵟、棕尾鵟、金雕、猎隼等，它们在草原食物链的顶端，捕食在草原生活的小型兽类（如鼠类）、鸟类及爬行类等。这是大鵟起飞瞬间。

This desert rangeland homes a wide variety of raptors, such as upland buzzards(*Buteo hemilasius*), brown-tailed buzzards, golden eagles, saker falcons and *etc*. Sitting on the very top of food chains of rangelands, these predators mainly prey on small animals (such as rodents), birds and reptiles that inhabit the grasslands. Shown in the photo is an upland buzzard that is just about to take on its wings.

2-3 极干旱荒漠草原生态系统

EXTREMELY DRY DESERT RANGELAND ECOSYSTEM

随着干旱的进一步加剧，干旱荒漠草原生态系统过渡到了极干旱荒漠草原生态系统，这是一类在极端干旱气候条件下形成的草地生态系统，在中国主要集中分布于本区，地形主要是盆地、山前洪积平原及沙地等。这里气候极端干旱，降水量特别稀少，部分地区甚至仅有数毫米到数十毫米。土壤为富含钙质与石膏的棕钙土、灰棕漠土，部分地区为盐渍化土壤。植被主要是旱生的小半灌木、半灌木和灌木。植物群落种类成分单一，分布稀疏。（灌木植被覆盖度超过40%的区域，往往也被划入森林生态系统的范围，见“一滴水生态摄影丛书”之一的《多样性的中国森林》，本书不再涉及。）

As the climate gets further drier, dry desert rangelands ecosystem will gradually be replaced by that of extremely dry desert rangelands, a type that normally takes shape in extremely dry climate conditions. Such systems in China exist mainly in this region in the basins, floodplains or sandy lands located at the foot of mountains. Weather here tends to be extremely arid, with annual precipitation ranging from a few millimeters to dozens millimeters. Soils found here chiefly belong to calcium- and gypsum-rich brown-calcium soils, gray-brown desert soils, and occasionally salinized soils in some places. Vegetation consists mainly of drought-resistant small scrubs, semi-shrubs and shrubs that are typically of monotonous species and sparsely distributed. (Areas where the shrubbery coverage exceeds 40% are often put under the category of forest ecosystems, for which this present book will not elaborate further. Readers interested can refer to *Diverse Forests of China* that has previously been published under the “A Single Drop of Water Ecological Photography Album series” .)

雪格布•新疆吐鲁番
Checked cloth of snow — Turpan, Xinjiang

在极干旱区，由于降水少、蒸发量大，荒漠草地植物的枝叶多呈尖刺状以减少水分蒸发，植株分布也呈现出生存空间分布最优的规律性布局。难得的一场白雪铺盖，将这种布局凸显了出来，看起来就像特意设计出的雪格布。

To better adapt to the precipitation-scarce and evaporation-high arid environments, plants grown over desert grasslands tend to bear spiky twigs and leaves that minimize evaporations. The distribution of plants also demonstrates regular patterns that fit well in with the environment and are of optimal layout. A just-in-time snowfall brings this optimal pattern into highlight, making it highly resemblant in look to checked cloth that has been intentionally-designed to this effect.

条条植物带·
甘肃阿克赛
Stripes of vegetation —
Aksai, Gansu

祁连山西端支脉的土尔根达坂山，山上白雪皑皑连连，山下干旱荒芜遍遍。广阔荒漠上一条条非常规则的平行线就像在图纸上用直尺画出来的一样，真让人不知究竟。其实登高一看，是荒漠大扇面上顺着一道道小水沟生长的草本植物，找到一个合适的角度观察，就能够看到图中的这类奇异景观。

The Turgan Dasan Mountain that sits at the west end of the Qilian Mountains is capped by thick white snow atop and surrounded by vast stretches of barren wilderness down at its foot. Over the vast wilderness, one will surprisingly see numerous parallel lines that occur in highly regular pattern, as if they had been drawn by someone deliberately. Only when you stand at a high place and look down, will you realize they are in fact just stripes of herbaceous plants that have sprouted up in the valleys that extend in parallel over the vast wilderness. You will be able to see this awesome view shown in the photo when the position you stand on is just right for the observation.

猛禽之王•新疆阿克苏

King of raptors —
Aksu, Xinjiang

金雕因其个大、凶猛、高居食物链顶端而被誉为"猛禽之王"，它是国家一级保护野生动物，且分布广泛，全中国范围内几乎都有它的分布。即便在极干旱荒漠草原，它也从不缺位。

Golden eagles(*Aquila chrysaetos*), for their huge sizes, ferocious disposition and top position on food chains, are reputed as the king of raptors. Listed among the Level I protected fauna species of China, they are extensively distributed at almost every part of China, including the extremely arid desert rangelands.

雪地柽柳•新疆库车
Tamarix chinensis in the snow — Kuqa, Xinjiang

图中多枝柽柳为灌木或小乔木，也称红柳，能耐烈日暴晒，耐干旱、耐水湿，又耐碱土，能在含盐量1%的重盐碱地上生长。在春、夏季节，骆驼乐食其嫩枝。多枝柽柳是中国西北干旱地区养驼业的重要饲料来源。

Seen in the photo are a few groves of *Tamarix ramosissima*, a shrubbery or small tree species that is also known as red willow. Highly tolerant against scorching sunlight, severe drought, excessive water and high salinity, this plant is capable of surviving salinized lands where saline-content reaches 1%. Camels are especially fond of eating their tender twigs during spring and summer seasons. For this reason, it is regarded as a very important source of fodders for camels in some arid regions of northwestern China.

荒漠草原雪•新疆奇台
Snow over the desert rangeland — Qitai, Xinjiang

这里是典型的荒漠草原，更是中国荒漠有蹄类野生动物分布最密集的地方之一。大雪降临，给大地披上了一层银装，为这里的野生动物提供了处处可以解渴的方便。

Shown in the photo is the most representative view of desert rangelands, which also happens to have the densest ungulate wildlife population across China. When heavy snow falls and covers the vast land up with a thick white blanket, it provides very convenient access to drinking water for wildlife that inhabit here.

荒漠草原植被与鼠兔•新疆巴里坤

Desert rangeland vegetation and pika — Barikun, Xinjiang

荒漠草原上，外形酷似兔子、身材和神态很像鼠类的鼠兔到处为窝。鼠兔是典型的草食性动物，有储草的习性。一只鼠兔可以破坏0.5～0.7平方米的草原。有的地方鼠害严重时，会造成严重的植被破坏、水土流失，导致草场退化，甚至出现严重的沙化现象。

Pika, an animal that resembles rabbits in appearance but mice in size and demeanor, burrows almost ubiquitously on desert rangelands. Being a typical herbivorous species that has the habit of storing grass, a pika can destroy 0.5 to 0.7 square meters of grassland on average. In places that are seriously threatened by pikas, severe vegetation destruction and soil erosion may occur, leading in result to grassland degradation and even grave sandification.

骆驼刺•新疆奇台

Alhagi camelorum — Qitai, Xinjiang

骆驼刺为半灌木，主要分布在中国内蒙古、甘肃和新疆等地，因其是荒漠草地及沙漠中骆驼四季都能够享用的草料而得名。骆驼刺在秋季结实和冬季枯黄后山羊和绵羊较喜食，春季萌发时牛喜食。骆驼刺庞大的根系能够在荒漠中聚成沙堆，在防风固沙方面具有非常重要的作用。

Alhagi camelorum, a semi-shrubbery species that are found to grow prevalently in desert grasslands and deserts in Inner Mongolia, Gansu and Xinjiang in China, is an important fodder that camels can feed on all the year around, hence often referred to as "camel thorns". Whereas goats and sheep like eating this plant when it is laden with seeds in autumn or when it withers in winter, cattle take it as their favored food only in spring when it begins to sprout. Its highly developed large root system can sometimes cluster together in deserts to form a huge sandy pile, thus playing a significant role in wind prevention and sand fixation.

白刺果·新疆喀什

Fruit of *Nitraria tangutorum* — Kashgar, Xinjiang

白刺属于灌木，分枝多而密集，呈丛生状，具有很强的固沙、阻沙能力，生长于荒漠和半荒漠的湖盆沙地、河流阶地及有风积沙的黏土地，是极干荒漠草原的代表性物种之一，其果是用途广泛的中药材。

Nitraria tangutorum, a dense-growing shrubbery that often exists in clumps, is particularly effective in fixing and tying down sands. One of representative plant species of extremely dry desert rangelands, it mostly grows at low-lying sandy places in deserts and semi-deserts, at terraced lands in river valleys, or at clay lands on which wind-borne-sands settle to accumulate. Its fruit is of pretty high medicinal value and is widely used in traditional Chinese medicine.

荒漠之舟•
新疆罗布泊
Ark of deserts — Lop Nor, Xinjiang

野骆驼是大型食草动物，又名野生双峰驼，仅生存在亚洲腹地的中国西北与蒙古国西南相邻的地方。全世界野骆驼的数量估计不超过1000只，远比大熊猫要稀少和珍贵。其生存环境极其恶劣，在荒凉、严寒、植被稀少、淡水奇缺的罗布泊荒原和库姆塔格沙漠，它们能够世世代代生存下来，可谓奇迹。

The wild camel(*Camelus ferus*), also known as Bactrian Camel, is a large herbivore that is indigenous only to the part of Asian hinterland that borders northwestern China and southwestern Mongolia. With its population in the wild estimated to be less than 1,000, Bactrian Camel is even rarer and more precious than giant pandas, particularly because it inhabits an environment that is extremely harsh and challenging. The survival of generation after generation of camels in the desolate, cold, plant-sparse and water-scarce Lop Nor Wilderness and the Kumtag Desert is nothing short of a miracle.

野放成功的野马•
新疆吉木萨尔
Wild horses successfully reintroduced back to nature — Jimushar, Xinjiang

这是世界上仅存的野马，在作为野马故乡的中国也已经灭绝。1985年，中国政府从英国引回了18匹在人工环境下生活的普氏野马，开始了野马回归自然的漫长工作并取得成功。图为放归在极干旱荒漠草原的野外环境中，自然繁育成功的野马群。

The Przewalski wild horses(*Equus ferus*), the only type of wild horses that still survives in the world, had long gone extinct in China — where they were indigenous. In 1985, the Chinese government, following the acquisition of 18 artificially-raised Przewalski wild horses from the United Kingdom, embarked on an arduous journey to reintroduce them back to natural environment, which turned out to be a very successful project. Captured in the photo are a herd of wild horses that are naturally bred in the wild after being reintroduced back to extremely dry desert rangelands.

2-4 草甸生态系统

MEADOW ECOSYSTEM

本区草甸生态系统主要分布在森林上线山地阳坡，与山地阴坡的高寒灌丛生态系统呈复合分布，被称为高寒草甸生态系统。在本区，高寒草甸生态系统多分布于天山、阿勒泰山等处，在一些地势低洼地区，还存在一定面积的盐碱土或盐化草甸土，这类土壤含盐量较高，比较湿润，或有短期的季节性积水。在这种条件下，只有比较耐盐的或盐生、泌盐的植物才能正常生长。这些植物主要是藜科、禾本科草本植物，但也常有一些零星的灌木，如柽柳、白刺等，它们形成的草甸生态系统被称为盐生草甸生态系统。

本区主要分布有高寒草甸生态系统、典型草甸生态系统、沼泽草甸生态系统和盐生草甸生态系统四类。

The meadow ecosystem in this region, which mainly exists on the sunny slopes of forested mountains, often occurs together with alpine scrub ecosystems found on the other side of mountains, hence often referred to as the alpine meadow ecosystem. Specifically speaking, the alpine meadow ecosystem in this region is mostly found in the Tianshan and Altai mountains. Some saline or salinized meadow soils that come with high salt content are also found in some low-lying areas that are relatively wet or subject to short-term seasonal waterlogging, where only salt-tolerant or salt-borne and salt-secreting plants can survive. These plants mainly include some herbaceous plants that fall within Chenopodiaceae and Gramineae families, but sparse-growing shrubberies like *Tamarix chinensis* and *Nitraria tangutorum* might also be present. The meadow ecosystem thus formed is generally known as the saline meadow ecosystem.

Four major types of meadows ecosystem exist in the region: alpine meadow ecosystem, typical meadow ecosystem, marshy meadow ecosystem, and saline meadow ecosystem.

高寒草甸羊亦肥•新疆和静
Fat sheep grazing over alpine meadows — Hejing, Xinjiang

巴音布鲁克地势平坦，水草丰盛，是典型高寒沼泽草甸生态系统分布区，更是新疆最重要的畜牧业基地之一。近二十多年来，草原植被覆盖面积总体上呈现增加趋势，除了人为破坏减少的因素外，还与降水和气温增高这两个重要因素有关，也与全球气候变暖的大趋势有关。

Bayinbuluk, with its flat terrain and abundant water and grass, is a region where the typical alpine marshy meadow ecosystem prevails. It also ranks top among the most important husbandry bases in Xinjiang. Over the past two decades, the vegetation coverage here has generally been on an increasing trend, which can probably be explained — besides by the decline of human-induced damages — by increased precipitation and warmer temperatures that are caused by the general trend of global warming.

阳光照耀着塔什库尔干·新疆喀什

Tashkurgan basked in the sun —
Kashgar, Xinjiang

帕米尔高原属严寒的大陆性高山气候，高原上有广阔的牧场，在塔什库尔干河谷地，分布着大面积的高寒草甸。帕米尔高原是中国的西疆极地，其平均海拔5000米以上，高原上90%以上的居民是塔吉克人。图为县城塔什库尔干镇附近的高寒草甸。

The Pamir Plateau, which falls under the harsh continental alpine climate zone, is covered by vast stretches of fine pasturelands, with extensive alpine meadows existing in the Tashkurgan valley. It is also the westernmost place of China, where the average altitude reaches over 5000m. Over 90% of the population living here are of Tajik nationality. Shown in the photo is a patch of alpine meadow that sits close to Tashkurgan Town.

干涸的草本沼泽·甘肃敦煌

Herbaceous marsh running dry —
Dunhuang, Gansu

库木塔格沙漠下的河滩绿洲迎来了秋天，河道已经干涸了，只有仍然生长旺盛的各种草本植物及龟裂的河底还在提醒你，这里的夏季曾经是充盈了河水的生命乐园。

Autumn is setting in over the riverine oasis tucked under the Kumutage Desert. With the river already dried up, and in spite of the crusted riverbeds, the rich variety of herbaceous plants that still flourish seem be reminding you that this site used to be teeming with vigorous lives back in water-rich summer times.

盐沼·甘肃敦煌

Saline swamp —
Dunhuang, Gansu

在干旱和极端干旱地区的一些地势低洼处，土壤含盐量较高又有季节性积水，只有一些比较耐盐的或盐生、泌盐的植物才能正常生长，形成盐生草甸生态系统。

In some low-lying areas that are relatively wet or subject to short-term seasonal waterlogging in arid and extremely arid places, only salt-tolerant or salt-borne and salt-secreting plants can survive, thus giving rise to some saline meadow ecosystems.

草甸秋色·
新疆阿勒泰
A meadow tinged with the color of autumn — Altay, Xinjiang

秋天降临在喀纳斯河谷的大片高寒草甸上，使草甸变得五彩斑斓，地势高一些的地方和山坡上的落叶松林似乎反应要慢了许多，仍然在留恋着夏日的时光，固守着，迟迟不愿脱掉绿色的衣装。

Autumn is tiptoeing in across the Kanas River valley, turning the vast stretch of alpine meadows there into a sea of variegated colors. In elevated parts and over the hill slopes, dense larches seem to be much slower in their response, as if still indulged in fond memories of the summer and reluctant to let go off their green attires.

盐角草•
新疆克拉玛依
Salicornia europaea —
Kramay, Xinjiang

盐角草耐盐、耐旱，亦耐水湿，是最耐盐的植物物种之一，多生长在盐碱地、盐湖旁及盐生草甸中。
Salicornia europaea, tolerant of salinity, drought as well as of excessive water, is a highly salt-resistant plant that may thrive on saline lands, saline meadows or by salt lakes.

六边形草圈•
新疆托克逊
Hexagonal patterns of grasses —
Toksun, Xinjiang

艾丁湖是吐鲁番盆地地表径流的归宿处，其洼地高程数据为海拔-154.44米，是我国陆地的最低点。湖周边的盐生草甸上出现了由不同的盐生植物构成的六边几何图形，非常神奇。
All surface runoffs in the Turpan Basin feed in the end into Lake Aydin, which, -154.44 meters in elevation, makes up the lowest part of China's land territory. On the saline meadows surrounding the lake, various saline-tolerant plants grow naturally in hexagonal patterns, presenting visitors with views that are almost magical.

2-5 可持续发展

SUSTAINABLE DEVELOPMENT

本区草原的过度放牧与开垦，药材与经济植物资源的过度采挖，水资源匮乏，开矿与工业及城镇发展带来的负面效应，给本已十分脆弱的干旱、极干旱地区生态系统带来许多严峻的挑战，例如，土地荒漠化严重、植物群落板块化，有毒、有害植物（相对于优良牧草而言，下同）相对比例上升，土壤退化，生物多样性丧失等。

本区为中国动物地理区划中古北界蒙新区的西半部分，为中温带和暖温带气候区中的干旱和极干旱地区。分布的野生动物以耐干旱的动物为主，包括中国特有的荒漠、山地野生动物。这里生活着很多有蹄类动物，如野骆驼、普氏野马、盘羊、蒙古野驴、岩羊和鹿类等大型食草动物，还有众多的过境鸟和留鸟，如鹫、隼、鹰、雕等猛禽和毛腿沙鸡等。草原植被主要由针茅、女蒿等禾本科、菊科等科的植物组成。

本区生态系统服务和生态功能的核心是防沙治沙、水土保持、草原畜牧业和生物多样性保护。干旱地区人类生存和发展最根本的约束因素是水，绿洲系统的规模和生态承载力更是由水资源的多少来确定的。因此，在实施措施上，应坚持生态优先，根据水资源和生态承载力确定利用方式和利用强度，合理调配流域内的水资源，控制人工绿洲规模，发展节水农业，提高水资源利用效益；严格限制放牧等生产活动，保护荒漠植被；加大力度修复严重沙化、风蚀、水蚀、水土流失、盐渍化的草地；将超过1/2的荒漠区纳入生态保护红线和自然保护地；发展可控的荒漠和绿洲旅游业；探索脆弱干旱荒漠草原生态保护与沙产业发展的新范式。

要把生态保护放在第一位，发展节水灌溉，大力提高水资源利用率；严格管控挖药材等破坏草原的行为；对地下煤炭、石油等矿产资源开采等破坏草原的后果，在谁破坏谁治理的基础上，要严格管控和加大奖罚力度。

在对土地进行综合统筹规划、生态影响评估和论证的基础上，可划出适当的地方发展风能、太阳能等新能源。同时，发展板下与板间畜牧业，提高和改善微生态环境，搞活经济并给牧区提供新能源。但应注意任意挤占和破坏草场的不良倾向。

经过长期的严酷的自然环境选择与人类的培育，在这一地区也形成了许多遗传性状稳定、品质优良的家畜品种，如新疆细毛羊、哈萨克羊等。也生存着许多具有特殊价值的珍稀野生动植物种，如野骆驼、野马、蒙古野驴、鹅喉羚等食草野生动物，以及雪莲、针茅、骆驼刺、肉苁蓉等野生植物。这里是中国难得的荒漠生物基因库。

在完善草原类自然保护地空间布局的基础上，要建设好天山、卡拉麦里、阿勒泰山、塔里木、巴丹吉林等草原类型和涉草类型国家公园，以达到建立和完善以国家公园为主体的干旱和极干旱地区自然保护地体系的战略目标。

Negative impacts caused by overgrazing and land reclamation, exploitative utilization of medicinal and economic herbal resources, severe scarcity of water resources, irrational development of mining industries, as well as by improper urban development all work together to present even more grave challenges to local ecosystems in arid and extremely arid regions that are already quite vulnerable. Take for instance, increasing land desertification,

一举多得·甘肃兰州
A multi-purpose initiative — Lanzhou, Gansu

在西北的戈壁、荒漠草原上，日照充足，是发展光电的理想地方。太阳能开发与荒地、荒漠治理有机结合，利用“板上发电、板下养羊、治荒改土”的“光伏+”绿色生态产业模式，既发展了绿色能源，又可以放牧养羊，改善了生态环境。

Sunlight-rich places across the Gobi and desert rangelands in northwestern China are ideal sites for generating photovoltaic electricity. Through the adoption of a holistic approach in developing solar energies and conserving wastelands and deserts, local communities have launched a multi-purpose program that combines power generating, sheep raising and soil upgrading into an integrated whole. The “photovoltaic + eco-industry” development model not only boosts the development of green energies, but also creates favorable conditions for developing sheep raising industry.

fragmentation of plant communities, incremental proportion of toxic or hazardous plant species (as against high quality fodders, applicable also to cases listed in the following parts), deteriorating soils, loss of biodiversity, all of these are leading in result to serious consequences.

In terms of China's zoogeographical zoning, this region belongs to the arid and extremely arid areas of the temperate and warm temperate climate zones, which in turn fall within the eastern part of the Inner Mongolia-Xinjiang Region of the palearctic realm. Wild fauna species inhabiting here belong chiefly to the drought-tolerant ones, including some that are indigenous to the desert and mountainous areas of China. In addition to the huge variety of ungulate herbivores species — like wild camels, Przewalski wild horses, *Ovis ammon*, Mongolian wild asses, *Pseudoisnayaur*, deer and *etc.* — that live here, this region is also home to countless migratory and resident birds, including both raptors such as buzzards, falcons eagles, vultures and snowy owls and prey species like Pallas's Sandgrouse and *etc.* Vegetation here mainly consists of plants belonging to the Gramineae and Asteraceae families, with *Stipa capillata, Ajaniatrifida* and *etc.* being a few examples.

The core eco-services provided by this region include desertification prevention and control, water/soil conservation, livestock husbandry and biodiversity conservation. A key restraint for the survival and development of human beings in arid places is water, whose availability plays a deterministic role in the scale of oasis systems and their ecological carrying capacity. In light of this objective condition, priority must be given to the needs of ecological conservation. Due considerations must be given to the availability of water resources and the ecological carrying capacity of the local area in the choice of resources utilization model and corresponding intensity. Reasonable allocation of water resources within the region shall have to be achieved through controlling the scale of man-made oasis, encouraging water-saving agriculture, and through improving the efficiency and effects of water utilization. Productive activities like animal grazing must be strictly controlled to protect the vegetation of deserts. Efforts must be strengthened in the restoration of grasslands that have been subjected to sandification, water/wind/soil erosion, or salinization. Over half of the deserts and wilderness in this region should be included within the system of protected lands and given prioritized protection. Development of desert- and oasis-based tourism must be scientifically guided so as to explore a new and innovative model that gives balanced attentions to the conservation of vulnerable and dry desert rangelands and the growth of sand-based industries.

Ecological conservation must be highlighted through encouraging water-saving irrigation techniques that contribute to improved efficiency in water resources utilization. Rangeland-damaging activities like herbal medicine digging must be put under strict regulation. With regard to the harmful consequences imposed on rangelands by coal and petrol-mining industries, we must, besides strictly adhering to the principle that holds the damage-causing party accountable for corresponding restorative efforts, adopt even more sophisticated regulating and punishing schemes.

Suitable sites, provided that well-coordinated planning, thorough eco-assessment and review have been done, can be identified as the designated areas for developing wind and solar energies. Meanwhile, animal husbandry industries can also be developed under or between photovoltaic panels so as to improve the micro-environment, boost local economic growth and new energy resources supplies. Nevertheless, attentions must be paid to preventing such undesirable consequences as irrational encroachment onto the rangelands or damages to them.

The long process of rigorous natural selection and painstaking artificial cultivation have also brought into existence a rich stock of genetically-stable and quality-assured livestock species that are well-adapted to the local conditions, and the beach sheep, Xinjiang Merinos and Kazakh sheep are just a few of the examples. Besides, this region is also home to some highly rare and precious wild flora and fauna species — the former includes Saussurea involucrata, Stipa capillata, *Alhagi camelorum*, *Cistanche deserticola* and *etc*., while the latter includes herbivores like wild camels, wild horses, Mongolian wild asses, Gazella subgutturosa and *etc*. — that make it a very unique and important gene-bank of desert biodiversity.

Besides further optimizing the spatial layout of rangeland-featuring natural protected areas, preparative efforts must also be made for the creation of a national park that covers the rich variety of rangelands existing in the Tianshan mountains, Karamali, Altay Mountains, Tarim and Badanjilin, so as to achieve the nation's strategic goal — putting in place and optimizing a national park-centered system of protected lands in arid and extremely-arid regions.

手中有草心不慌•新疆白哈巴

Abundant hays supply keeps worries at bay — Baihaba, Xinjiang

白哈巴村以图瓦人为主，被誉为“西北第一村”。这里一个非常重要的景观是每家每户的草垛子，这些草料都堆在架空的木架子上，堆得有模有样、整整齐齐。这是冬季来临之前，图瓦人为了牲畜过冬而精心准备的草料堆。

Baihaba Village, a community of the Tuva people, is credited as the "No. 1 village in Northwestern China". A most impressive view in this village is the neatly stacked hay piles that are put on wooden shelves in front of each household. These are the fodders that the Tuva people have carefully prepared for their livestock prior to the advent of winter.

大转场•新疆阿勒泰

Move between seasonal pastures — Altay, Xinjiang

这也许是世界上规模最大的转场。每年季节变换，哈萨克牧民就用这种四季迁徙的独特生产方式，在不同草场之间转移，这使草场得以休养生息，这种可持续的生产方式维护着该地区人与自然的生态平衡。

This presumably is the most impressive and largest-scaled move of livestock in the world. As the seasons evolve, Kazakh herdsmen would move regularly between their seasonal pasturelands so as to allow times for the grasses to recover and regenerate. This time-honored sustainable means of production that exists uniquely here is highly conducive to the ecological balance between people and nature.

basin rangelands — demonstrate notable characteristics that are typical of vertically-shaped belts. The elevation at which this type of ecosystem occurs varies depending on the height at which snowlines and glaciers begin to show on each mountain as well as on how steep the particular slope is. Generally speaking, the point at which this occurs tends to decrease as we move from the south to the north, being 5,800m, 4,700m and 3,800m in altitude at the Kailas Range, the Bayan Har Mountains, and the Tianshan Mountains respectively. The screes normally border perm-frosted belts at the top and extend downwards until the belts of alpine cushion plants, alpine rangelands (in cases where the mountain slopes are especially steep, the scree may slide further downwards by 200–500 meters), even further down to alpine meadows and alpine shrubberies. At mountain tops where the slopes are comparatively gentle, they may occur alternatively with alpine meadows and alpine cushion plants.

The occurrence of sparse vegetation on screes is the result of continuous elevation of high mountains that are liable to be stricken by multiple rain- and snow-falls or that of hailstones in a single day. Under such an extremely harsh natural environment, only plants that come with extraordinary ecological and biological properties will be able to survive and flourish, hence constituting the pioneering plant communities on the screes atop high mountains. With the annual average temperature of the habitat standing at minus 4 Celsius degrees and that of the warmest month also below 0 Celsius degrees — which could be made even worse by notably high fluctuation between day-night temperatures, strong solar radiation and relentless winds, plants growing here tend to demonstrate some features that are typical of alpine condition-fitting species. Specifically speaking, they tend generally to be small in size, shaped like lotus pedestals or cushions, covered in leaf and stem by dense wooly hairs, and have highly developed root systems. To catch the attentions of insects which are essential for their propagation, their flowers are often large in size, bright in color and long in blooming span, representative examples including *Meconopsis* spp., *Saussurea involucrata*, *Rheum nobile*, *Lagotis* spp. (also known as *Ligotidis* spp.) and *etc*..

马尿泡•青海天骏

Przewalskia tangutica —
Tianjun, Qinghai

马尿泡是多年生草本植物，多生于海拔3200~5000米的高山砂砾地及干旱草原上，因果萼呈椭圆状或卵状而得名，是诸多高山流石滩植被生态系统的植物物种组成之一，也是藏药材之一。

Przewalskia tangutica, a perennial herbaceous plant that typically grows at the 3,200–5,000m-high gravel land and arid rangelands, is named in Chinese as *Maniaopao* for its oval- or egg-shaped fruit. In addition to being a major composing plant of the alpine scree ecosystem, it is also a very important Tibetan medicinal herb.

高山冰缘带植被·甘肃阿克塞

Alpine periglacial plants —
Aksai, Gansu

这是以大雪山（其主峰海拔5483米）为背景的高山冰缘带植被景观，高山流石滩上各种苔藓、垫状植被等呈点状稀疏分布。

This is the landscape of alpine periglacial vegetation, looming in the background is the 5,483m-high Daxueshan. Sparsely scattered across the alpine screes are a variety of mosses and cushion plants.

地衣和雪·西藏林芝

Lichens over the snow —
Nyingchi, Tibet

地衣可分为壳状、叶状和枝状三种类型，多喜光，耐寒和耐旱性都很强，但不耐大气污染，可以在高山带、冻土带等其他植物不能生存的地方生存。地衣所分泌的多种地衣酸可腐蚀岩面，使岩石表面逐渐龟裂和破碎，加之自然的风化作用，逐渐在岩石表面形成了土壤层，为其他植物的生长创造了条件。因此，地衣常被称为"植物的拓荒者"。

Lichens, a light-prone plant that mainly appear in shell-like, leaf-like and branch-like forms, are especially cold- and drought-tolerant but sensitive to air pollutions, and therefore can survive alpine frosted areas that most other plants cannot. Lichens can secrete a variety of corrosive lichen acid that makes rock surface crack and break into smaller pieces, which, when further exposed to the wear and tear of natural environment, would in turn form a soil layer that provides a fairly favorable condition for other plants to survive. For this reason, lichens are sometimes called the "pioneer species among plants".

流石滩下牦牛群•青海久治

Yak herds grazing below the alpine screes line —
Jiuzhi, Qinghai

利用航拍，阿尼玛卿雪山的高山冰缘带下，冰川流水、高山流石滩和诸多冰斗湖都一一展现在了我们的面前。在如此贫瘠粗糙的流石坡面上，草原生态系统都没有缺位，只要有点条件，小草就无处不在。仔细观察，我们还能看到图最下方的湖边、草地上星星点点散布着正在食草喝水的牦牛群。

With the help of aerial photographing techniques, we are able to catch sight of the various views that exist below the periglacial belt on the Animachen Snow Mountains — meltwater from glaciers, alpine screes and numerous tarns. Even on this barren and coarse slope made of alpine screes, rangeland ecosystems can still be found to exist. If only there is the slightest chance of survival, grasses will manage to sprout and flourish. A careful look at the photo will reveal that, over the grassland that stretches out around the lake seen at the bottom part, a herd of yaks are leisurely grazing and drinking.

总状绿绒蒿•青海果洛

Meconopsis racemose —
Golo, Qinghai

在海拔4500米的高山流石滩上，笔者终于找到了一个石坎，利用强烈的阳光和石坎的阴影，成功地拍摄了一张低角度暗背景的总状绿绒蒿全株照片。逆光效果让绿绒蒿最有特性的绒毛毕现，正在盛开的花朵和未开的花蕾齐全，全株上下颜色鲜艳，立体感强，旁边还有一株已开谢的花作对比虚化陪衬。这是笔者趴地上，大口喘气、相机持不稳的状况下拍摄的，也是笔者迄今为止最满意的一张绿绒蒿照片。

I was finally able to, at the 4,500m-high alpine scree, find a stone dam, where he could take advantage of the strong sunlight and the shades created by the dam and managed to capture a complete view of a full crop of *Meconopsis racemosa*. Highlighted by the comparatively dark background, the most defining feature of this plant — its fluffy hairs — is depicted to the fullest effects. Both flowers in full bloom and budding ones are all captured to present a vivid and lively view, which is made all the more eye-appealing by a fading flower in the background. This photo, though taken under a most awkward condition where I had to spread myself out at full length on the ground and could hardly keep my camera steady for the shooting, turns out to be one that I am so far most satisfied with among all the pictures I have ever taken that features this plant.

唐古特雪莲·青海果洛

Saussurea tangutica — Golo, Qinghai

雪莲生长的环境极其恶劣，在那里一般的植物根本无法存活，而雪莲却能在零下几十摄氏度的严寒中和空气稀薄的缺氧环境中傲霜斗雪、顽强生长。图中的唐古特雪莲生长在海拔3600~4800米的风化带和雪线上的石隙、砾石中，能迎着寒风、冰雪及烈日毅然开放，生命力极强。图为处于盛花期的唐古特雪莲。

Saussurea tangutica grows in an extremely harsh environment where almost no other plant can manage to survive. In spite of the minus several ten Celsius degree temperature and the highly oxygen-deficit thin air, this plant manages to survive and thrive. Shown in the photo is a full-blooming *Saussurea tangutica* flower that was found to exist at a rocky crevice above the 3,600–4,800m snowline. The fact that it was able to survive and thrive in spite of the biting coldness, the icy snows and merciless sunlight testifies convincingly to the vitality of life. Shown in the photo are *Saussurea tangutica* in full blossom.

3-2 高寒垫状植被生态系统

ALPINE CUSHION VEGETATION ECOSYSTEM

高山垫状植被同样是一类特殊的植被类型，是由耐寒抗旱的垫状半灌木、小灌木及草本植被组成。垫状的生长形式既是高山严酷的水热条件、辐射和强风等对植物“塑造”的结果，又是植物经历高寒生境长期自然选择演化形成的。同样是高山冰缘带植被的主要组成，这类草本或小灌木的小枝生长极度受抑制而呈辐射密集分枝，形成半球形或凸起的垫状体，其叶小、质硬而密，覆于表面，抗强风，常呈狭窄的带状或块状分布。为了传代引起昆虫注意，花朵颜色鲜艳且大，花期较长。由于垫状植物的新陈代谢和生长过程十分缓慢，它们的寿命很长，可达百年以上。

The alpine cushion vegetation is also a very special type, composed mainly of cold-resistant and drought-tolerant cushion semi-shrub, small shrubs and other herbaceous plants. Cushion plants are the result of both external shaping and internal adaptation, the former referring to the “modelling” of plants by the harsh natural elements in high-mountain environment (including water, heat, solar radiation and winds), whereas the latter referring to the natural adaptation of plants that aims to better fit the harsh alpine environments. Also a major composing part of the periglacial vegetation belt, these herbaceous or small shrubs tend to have tiny twigs that, due to the strong restricts their growth is subjected to, spread densely out to resemble a semi-sphere or undulated cushions. Their leaves tend to be small, hard and dense, clinging tightly together to narrow strips or patches so as to make themselves more resistant against strong winds. To attract insects that are crucial to their propagation, their flowers tend to be bright in color, large in size and long in blooming span. Because of their extremely slow metabolism and growing process, most cushion plants tend to have very long lifespan, lasting often for several hundred years.

垫状植被及野牦牛•青海格尔木
Cushion vegetation and wild yaks — Golmud, Qinghai

垫状植被是高山冰缘带的主要植被组成之一，这类草本或小灌木的小枝生长受极度抑制而呈辐射密集分枝，形成半球形或凸起的垫状体，像一层层、一块块厚厚的地毯铺垫在高山坡上。远处一头野牦牛警惕地看着镜头，即将竖起的尾巴表明它已经开始发怒了……

The cushion vegetation is a major composing part of the periglacial belt. The herbaceous or small shrubs tend to have tiny twigs that, due to the strong restricts their growth is subjected to, spread densely out to resemble semi-sphere or undulated cushions that blanket high-lying mountain slopes layer after layer. The wild yak(*Bos mutus*) in the distance is looking vigilantly at my camera, the high-rising tail suggests that it is getting increasingly infuriated …

海子山下•四川甘孜

At the foot of the Haizi Mountain — Ganzi, Sichuan

海子山为雅砻江和金沙江的分水岭，在有限的地域里集中了大大小小1000多个海子（湖泊），其规模、密度在中国是独一无二的，故名海子山。这里平均海拔4410米，有特征明显、面积很大的垫状植被在此分布。

Being the watershed between the Yalongjiang River and the Jinshajiang River, the narrow strip of Haizi Mountain is dotted with over 1,000 variously-sized lakes, which is matchless in China in terms of both the scale and density, hence named as such. Averaging 4,410 meters in elevation, this area is known for its extensive coverage of distinctive and sprawling cushion vegetation.

▲ 垫状草地遍羊群·西藏日喀则

A cushion grassland teeming with sheep — Shigatse, Tibet

在高海拔的雪山之下，草本植被呈半球形或凸起的垫状体，叶小、质硬而密，覆于表土面，能抗强风。这里虽然自然条件严酷，植被生物量不高，但依然是牛、羊生存所依赖的草原宝地。

Below the tall snow-capped mountains, the herbaceous plants tend to appear in the form of semi-sphere or undulated cushions. Their leaves tend to be small, hard and dense, clinging tightly together to narrow strips or patches so as to make themselves more resistant against strong winds. In spite of the harsh natural environment and low biomass, these areas still make up precious rangeland resources on which cattle and sheep depend for sufficient fodder supplies.

▼ 独一味·西藏桑日

Phlomoides rotata — Sangri, Tibet

草本植物独一味，生长于海拔2700~4900米的高原或高山上强度风化的碎石滩中或石质高山草甸、河滩地中。其贴地生长的习性导致在有灌丛的情况下不易存活。在中国民间，独一味全草可入药。

Phlomoides rotata is a herbaceous plant that mainly grows on plateau, highly wind-eroded gravel lands, or rocky alpine meadows and river flats that lie at the 2,700–4,900m altitude-range. It tends to cling to the ground, which makes it less viable in places where shrubberies are present. The entire crop of *Phlomoides rotata* is of high medicinal value in traditional Chinese medicine.

垫状植被·甘肃阿克塞

Cushion plants —
Aksai, Gansu

由于昼夜温差大、辐射强、风力劲，垫状植被适应高寒环境的形态特征而颇具特色。垫状植被植株矮小、莲座状或垫状，植物体密被绵毛，植物根系非常发达。为了吸引昆虫传播花粉，垫状植被花朵较大，且颜色鲜艳、花期较长。

To better adapt to the alpine environment marked by large day-night temperature variation, strong solar radiation and strong wind, cushion plants have developed their distinctive biological and morphological features. They tend generally to be small in size, shaped like lotus pedestals or cushions, covered in leaf and stem by dense wooly hairs, and have highly developed root systems. To catch the attentions of insects which are essential for their propagation, their flowers are often large in size, bright in color and long in blooming span.

棕熊一家出游·
青海可可西里
An outing day for the brown bear family — Hoh Xil, Qinghai

在高寒的草原带上，母棕熊带着三头小熊趁天快黑的功夫闯进一处工地来寻找食物，一头藏獒的狂吠引起了笔者的注意——笔者观察到母熊正在踌躇不前，一头小熊开始要另辟蹊径走下石坡了。棕熊是杂食性动物，主要采食各种草及果实等植物性食物，偶尔摄食动物性食物。青藏高原的棕熊，从荒漠边缘至高寒草原，甚至冰缘地带，都能顽强地生活。

As darkness gathers in on the alpine rangelands, a female brown bear(*Ursus arctos*) and her three young cubs break unexpectedly into a construction site to look for some food. The ferocious barks of a Tibetan mastiff catch my attention. It is at that momentthat I notice the hesitant mother bear and a young cub making its way towards the lower part of the rocky slope. An omnivorous animal species, brown bears mainly feed on grasses and fruits of plants, and only occasionally hunt other animals for food. Brown bears inhabiting the QTP have their presences in a wide range of places — from the edge of deserts to alpine rangelands, even to periglacial belts.

绿尾虹雉·
四川阿坝
Chinese Impeyan — Aba, Sichuan

绿尾虹雉为中国特有种，尤其喜欢在海拔3000~5000米的多陡崖和岩石的高山灌丛和灌丛草甸中活动，靠挖掘植物的根、地下茎等地下部分为食。雄性绿尾虹雉羽衣华丽，具金属光泽，数量极为稀少，故被称为中国的“国鸟皇后”。

Chinese Impeyan(*Lophophorus lhuysii*) , an indigenous species that exists only in China, is particularly widely distributed in alpine shrubberies and shrub meadows located at 3,000–5,000m-high steep slopes and cliff, feeding primarily on the roots and under-buried stems of plants. Male Chinese Impeyan is covered with colorful and splendid plumage that gives off a metallic luster. Because its population in the wild is already extremely small, this species of bird is sometimes hailed as the “queen of national bird” of China.

3-3 高寒草甸生态系统

ALPINE MEADOW ECOSYSTEM

高寒草甸生态系统分布在青藏高原东部、中部和高原东南缘高山及祁连山等海拔3200~5200米的地带，是青藏高原区分布最广的草原类型。在森林上线山地阳坡，与山地阴坡的高寒灌丛呈复合分布。其分布区气候特点是：高寒、中湿、日照充足、太阳辐射强、风大。在青藏高原的东部，夏季受东南季风与西南季风的强烈影响，水汽充足，湿润多雨，即使最暖的月份也经常出现霜冻和雪，年降水量400~700毫米，多集中在6~9月。这些草甸生态系统共同的特征是所处地区海拔都比较高，气候寒冷，死亡植物分解缓慢。

在青藏高原大气环流控制下，降水自东南向西北逐渐减少，与此相适应的高寒草甸构成了青藏高原东部的高原地带性植被的重要组成。它的分布下限可延伸到高寒灌丛带，与山地森林带、山地草原相衔接。其分布上线与高寒垫状植被、高山流石滩稀疏植被相衔接，甚至有些种类可加入后二者当中，成为它们的重要组成成分。

在严寒气候的影响下，高寒草甸植物种类组成相对较少，却是高寒草地生态系统种类组成最丰富的类型，具有草群生长密集、覆盖度大、生长季节短和外貌整齐的特点。

组成高寒草甸的多数植物长期适应高寒气候而具有丛生、莲座状、株矮、叶小、被茸毛、生长期短和营养繁殖等特性。在青藏高原从东（南）到西（北），气候逐渐变干，种类组成由多变少，草层由高变低。

在本区，高寒草甸生态系统主要包括高寒典型草甸生态系统、高寒沼泽草甸生态系统和高寒盐生草甸生态系统，在一些沟谷和水分含量较大的地方，还存在有条块状的高寒灌丛草甸生态系统。

Alpine meadow ecosystem is distributed in the eastern and central parts of the QTP, the high mountains at its southeastern edge and the Qilian mountains, where the altitude mostly ranges between 3,200–5,200 meters. It may also exist above the forest line on the sunny slopes or side by side with alpine scrub on the shady slopes of tall mountains. The defining characteristics of the climate in areas where this ecosystem exists include the following: being cold, moderately wet, sunlight-rich, windy, and of strong solar radiation. Due to the strong influences of the southeast and the southwest monsoons, the eastern part of the QTP is endowed with plenty of moisture, making it fairly wet and rainy. Snow and frost may strike even in the warmest months, with precipitation — averaging 400–700 millimeters annually — falling mainly during the period ranging between June and September. A common feature shared by all these meadow ecosystems is that they are all high in altitude, cold in climate, and slow in the decomposition of dead plants.

Due to the influences of the prevalent atmospheric circulation on the QTP, precipitation gradually decreases as we move from the southeast to the northwest. As a result, alpine meadows, which adapt well to this condition, become the dominant vegetation of the mountainous area in eastern QTP. The lower boundary of this ecosystem may extend downwards until the alpine scrub belt that borders on the mountainous forest belts and mountainous rangelands. Its

泽曲草原•青海久治

Zequ rangelands — Jiuzhi, Qinghai

泽曲草原位于青藏高原东部的河谷地区，这里地势宽阔平坦，平均海拔3700米左右，泽曲河属黄河一级支流，有大小30多条支流，弯弯曲曲像毛细血管一样分布在地表上，形成并滋养了大片的高寒草甸。

The Zequ rangeland sits in the valleys of eastern QTP, where the terrain tends to be open and flat, averaging 3,700m in altitude. Besides the Zequ River — a primary tributary of the Yellow River, over 30 other tributaries of various sizes meander across the vast territory, nurturing behind them vast stretches of alpine meadows.

upper boundary borders with the alpine cushion vegetation and the sparse vegetation of alpine scree, with certain species even extending further into the latter two ecosystems and forming an important component of them.

Under the influences of the frigid and harsh climate, alpine meadows are typically composed of fewer plant. Nevertheless, they still make up the type that has the richest plant diversity among all alpine grassland ecosystems, demonstrating the following characteristics: densely-growing grass communities, relatively high coverage, short growing season, and comparatively neat appearance.

Most plants species that make up the alpine meadows have over the history developed some very distinctive features that adapt well to the alpine climate. Specifically speaking, their small and lotus pedestal-shaped crops tend to grow in clumps; their leaves tend to be small and covered by dense velvet hairs; they tend to have short life spans, and typically reproduce through nutritional organs. As the climate on the QTP gets increasingly drier from the east (south) to the west (north), the composing species also start to decline in diversity, and grasses become increasingly shorter in crop height.

The alpine meadow ecosystems of this region mainly include the following types: alpine typical meadow ecosystem, alpine marshy meadows ecosystem, and alpine saline meadow ecosystem. Stripes of alpine shrubbery meadows can also be seen in some low-lying valleys or water-rich places.

高寒草甸•
四川稻城
Alpine meadows — Daocheng, Sichuan

位于横断山区的四川亚丁国家级自然保护区有一片高寒草甸和沼泽草甸组成的草地，成群的野生岩羊和人工放牧的牛、羊在这里共同享受着充足的阳光、青青的草地和纯净的河水。图中远处的雪山为央迈勇雪山。

Lying in the Yading State Level Nature Reserve in Sichuan is a vast stretch of grasslands that is consisted of both alpine meadows and marshy meadows, where herd upon herd of wild *Pseudoisnayaur* and domesticated cattle and sheep bath leisurely in warm sunlight, enjoying palatable green grasses and pure water. Looming in the distance in the photo is the Yangmaiyong snow mountain.

松赞林寺下•
云南香格里拉
By the side of the Songzanlin Lama Temple — Shangri-La, Yunnan

云南西北部的横断山区是青藏高原的东南延伸，这里水、热条件比较好，平坦低洼的地方形成了大片的高寒草甸生态系统。图为著名的松赞林寺下大片的高寒草甸、高寒沼泽草甸（此处海拔3300米）。

The south-eastern part of the QTP extends until it reaches the Hengduan Mountains situated in north-western Yunnan Province, where favorable combination of water and heat gives rise to vast patches of alpine meadow ecosystem in some low-lying places. Shown in the photo is a vast stretch of alpine meadows and alpine marshy meadows lying next to the highly reputed Songzanlin Lama Temple (3,300m in altitude).

秋色藏野驴群•
青海玛多
A herd of wild Tibetan asses under the autumn sky — Maduo, Qinghai

在黄河源区的高寒盐生草甸上，藏野驴们正在努力吃草“抓膘”，以迎接即将到来的漫漫寒冬。图中红色的是西伯利亚蓼，黄色的是青藏薹草。

Over the alpine saline meadows from where the mighty Yellow River has its source, wild Tibetan asses(*Equus kiang*) are busy filling their stomach to the fullest so as to get sufficiently prepared for the long and harsh winter that will soon set in. The red part in the photo is covered by *Knorringia sibirica*, and the yellow part by Qinghai-Tibetan *Carex* sp.

哈拉湖畔•
青海天骏
At the bank of Lake Harrah — Tianjun, Qinghai

属于无人区的哈拉湖是一个内流湖，是青海省仅次于青海湖的第二大湖，湖畔有大片的高寒盐生草甸。其中，红色的是西伯利亚蓼。远处的雪山是疏勒南山，最高峰岗则吾结（海拔5808米）是祁连山脉的最高峰，是疏勒河上游谷地与哈拉湖盆地两内流水系分水岭的最高点。

Lake Harrah, an endorheic lake situated in anuninhabited region, is the second largest lake in Qinghai province — next only to the Qinghai Lake. It is surrounded by vast patches of alpine saline meadows dominated by red *Knorringia sibirica*. The snow-capped mountain looming in the distance is the south Shule South Mountain, whose peak — the 5,808m-high Gangzhegujie — marks the highest point of the entire Qilian Mountains as well as the highest point of the watershed that separates the upper reaches of the Shule River and the basin of Lake Harrah.

烟雾高寒草甸·
西藏那曲
Alpine meadows enshrouded in dense mists — Nagchu, Tibet

那曲位于西藏北部的唐古拉山脉、念青唐古拉山脉和冈底斯山脉之间，平均海拔5200米以上，是西藏气候条件最恶劣的地区之一。这里高寒缺氧，气候干燥，昼夜温差大，多大风天气。那曲草甸以辽阔、高寒著称，主要由小嵩草组成，且生长密集，由于人为干扰少，自然状态保存最好。图为处处腾着烟雾寒气的那曲草甸，远处为青藏公路和上面行驶着的大货车。

Situated between the Tanggula Mountains, the Nyenchen Tanglha Mountains and the Kailas Range in northern Tibet, the 5,200m-tall Nagchu ranks atop among places in Tibet that have the harshest climate conditions for its oxygen-deficit air, dry weather, sharp day-night temperature fluctuation and frequent occurrence of gales. The vast and cold Nagchu Meadows, which are predominantly covered by dense-growing *Kobresia humilis*, are preserved in an almost intact natural condition that is free from human interference. Shown in the photo is the Nagchu meadows enshrouded in dense mists, with the Qinghai-Tibetan Road and its trucks traffic looming vaguely in the distance.

西藏杓兰·西藏林芝

Cypripedium tibeticum —
Nyingchi, Tibet

兰花一般生长在深山幽谷的山腰谷壁，透水和保水性良好的倾斜山坡或石隙，以及稀疏的山草旁。

Orchids generally grow on slopes or rock crevices in remote and secluded mountainous valleys that come with soil that has good water permeability and retention, or grow side by side with sparse grasses.

蓝玉簪龙胆·青海玉树

Gentiana veitchiorum —
Yushu, Qinghai

蓝玉簪龙胆生长于海拔2500~4800米的高山和亚高山地区，见于高山草甸、灌丛、山坡草地等。蓝玉簪龙胆是草地上最迟开的花之一，秋冬时节在一片枯黄的草丛中，蓝玉簪龙胆开花一片片一簇簇，临风摇曳，显出一种淡雅、素静的美。

Gentiana veitchiorum is typically found on alpine meadows, shrubberies or grasslands that exist in the 2,500–4,800m-high mountainous and sub-mountainous regions. A late-blooming species, clusters of this plant is often found to bloom and dance gracefully among withered grasses in late autumn and early winter, presenting an almost indescribable beauty of elegance.

玛曲草原·甘肃甘南

Maqu rangelands —
Gannan, Gansu

在广阔肥美的玛曲草原上，河曲缓缓流过，数量庞大的牦牛群在草甸与河曲之间自由地来回游荡。这里是甘肃最富饶的大草原之一，同时也是甘肃畜牧业发展的最重要支柱之一。

Over the vast and lush Maqu rangelands where the Hequ meanders quietly across, huge herds of yaks roam leisurely on meadows filled with highly palatable grasses. Ranked among the most fertile rangelands of Gansu Province, this area constitutes a crucial pillar on which the husbandry industry of the province depends.

▲ 党河源头•甘肃肃北

The source of the Danghe River — Subei, Gansu

党河是敦煌人民的母亲河，它从哈拉湖附近发源，在党河南山和野马南山之间宽阔的河谷中穿行。这是党河源头附近高寒草甸上的牦牛群。

The Danghe River, the mother river of people in Dunhuang, has its source near Lake Harrah and flows quietly by the wide valley between the Danghe South Mountain and the Yema South Mountain. Shown in the photo is a herd of yak grazing over the alpine meadows near the source of the Danghe River.

▼ 草原鹤鸣•四川若尔盖

Songs of cranes over the rangelands — Ruoergai, Sichuan

水草丰美的若尔盖草原是四川省最大的草原，是全国最好的天然牧场之一，由高寒草甸和沼泽草甸组成。藏族牧民世代在这里逐水草而居，与他们的牦牛、绵羊、马，包括黑颈鹤等鸟类在一起，过着自由自在的游牧生活。这里是全球黑颈鹤最大的越冬地、繁殖地。

The water-abundant and grass-lush Ruoergai rangelands, the largest among its counterparts in Sichuan Province and one of the best natural pasturelands in China, is mostly composed of alpine meadows and marshy meadows. For thousands of years, generation after generation of Tibetan herdsmen — together with their yaks, sheep, horses as well as black-neck crane(*Grus nigricollis*) and other fowl species — have been living carefree nomadic husbandry-based lives here. This place is also the world's largest site where black-neck cranes overwinter and breed.

清晨的高原鼠兔·青海都兰

Pika in early morning —
Dulan, Qinghai

尽管是夏季，湿气还是让清晨的高寒草甸霜冻遍地。高原鼠兔从洞里出来探望，为的是迎接草原上最早的能给它及时带来温暖的阳光。

Despite it is summer, the highly humid air over the alpine meadows still makes the rangelands in early morning covered with a thick layer of frosts. The pika(*Ochotona curzoniae*) is peeping out from its hole, as if eager to usher in the first beam of sunshine over the rangelands that can hopefully bring warmth to him.

红花绿绒蒿·青海班玛

Meconopsis punicea —
Banma, Gansu

绿绒蒿被誉为“离天空最近的花”，花开艳丽、花形大，看似弱不禁风，其实不畏严寒，能在流石滩中昂首不倒。图为迎风招展的红花绿绒蒿。

Credited as “the flower that stands the closest to the sky”, blossoms of the *Meconopsis* are bright in color and large in size. Its seemingly vulnerable and delicate crop can in fact survive the harsh environment of biting cold and barren screes. Shown in the photo is *Meconopsis punicea* that is dancing gracefully in the breeze.

隐蕊杜鹃·西藏工布江达

Rhododendron intricatum —
Gongbo'gyamda, Tibet

隐蕊杜鹃为常绿灌木，高0.15~1米。分枝密集而缠结，生长于高海拔地区，喜凉爽湿润的气候，恶酷热干燥，在高山草甸中常常呈散点或群丛分布，花朵美丽，颜色鲜艳。

Rhododendron intricatum is an ever-green shrub, generally being 0.15–1m tall. This highly vibrant rhododendron, with its dense twigs, verdant leaves and graceful blossom that come in a rich variety of color, grow mostly in high altitude areas where the climate is cool and humid, since it is very aversive to dry and hot weather. They can often be seen growing in clusters or sparsely dotted amidst alpine meadows.

高寒植被与牦牛·青海都兰

Alpine vegetation and yak —
Dulan, Qinghai

这是在高海拔的一个阴坡沟边。特殊的小环境中，高寒植被生长比较茂盛，许多株红花绿绒蒿绽放在金露梅的矮灌木丛中，一头牦牛正边吃草边散步其中。也许，此时牦牛的肚子里已经装进去不少叶嫩汁多的国家二级保护野生植物——红花绿绒蒿了。

Shown in the photo is the view of a shady high-altitude valley. Among the lush alpine vegetation thriving in this unique niche, numerous crops of *Meconopsis punicea* are blooming brightly amidst the low *Dasiphora fruticose* shrubberies by which a yak is foraging leisurely. Presumably, the yak's belly is already filled half full with the tender and juicy *Meconopsis punicea*, a Level II protected wild plant species of the country.

3-4 高寒草原生态系统

ALPINE RANGELAND ECOSYSTEM

高寒草原是指以耐寒的多年生旱生禾草、薹草和小半灌木等植物群落构成的草原，具有草群稀疏、覆盖度小、草丛低矮、层次结构简单的特点。群落中伴生着适应高寒干旱生境特殊条件的垫状植物和许多高山植物种类，其生长季短、生物生产量低。

高寒草原主要分布在青藏高原面上的中部高山，是在大陆性气候强烈、寒冷而干旱的地区所特有的一个草原生态类型。在高原面上，高寒草原分布幅度较为宽广，具有高原地带性分布特征。

Alpine rangelands refer to plant communities that consist mainly of cold- and drought-resistant perennial herbaceous plants, *Carex* spp. and small semi-shrubs, presenting such characteristics as being sparse in growth, low in coverage, short in tussock height, and simple in layer structure. The plant communities are occasionally strewn with cushion plants and alpine species whose short growing season and low biomass adapt well to high-lying cold and arid habitats.

Alpine rangelands, a type of rangeland ecosystem unique to cold and arid regions that are subjected to the strong influences of continental climate, exist primarily on the QTP. Alpine rangelands are extensively distributed over plateaus and tend to demonstrate obvious horizontal patterns in their geographical positions.

石林草地羊群·青海天骏
Sheep herds over the Shilin grasslands — Tianjun, Qinghai

天峻县地处青藏高原东北部、祁连山南麓、柴达木盆地东部。天骏石林片区中高寒草原与裸石相间，羊群点缀在绿色的草地之中，颇有人间天庭之感。

Tianjun County is tucked between the northeastern part of the QTP, the southern foot of the Qilian Mountains and the eastern part of the Qaidam Basin. In the Shilin area where alpine rangelands exist side by side with bare rocks, herds of sheep are often found to dot randomly across the green grasslands, at the sight of which one will can't help but feel as if having arrived at a heavenly place on the earth.

天高云淡看草原•
青海门源
Rangelands under the blue sky and white clouds — Menyuan, Qinghai

祁连山脉位于青藏高原的北部，由于东西走向的祁连山对南方季风带来的湿润空气和雨水的阻隔和保留，在青海的互助、门源、祁连等县域内形成了一条水草丰茂的草原带。祁连山草原是中国最美的草原之一，山清水秀景如画，天高云淡夏如春。祁连山脉中涵养并不断流出的水源是持续滋养这片草原永不疲倦的推动力。

The east-west going Qilian Mountains at the northern QTP forms a blockade against the humid air and rains that come with the southernly monsoon, leaving behind it a vast belt of flourishing rangelands at such counties as Huzhu, Menyuan, Qilian and *etc*. The rangelands at the Qilian Mountains rank among the most beautiful rangelands across China, presenting viewers with a picturesque landscape made of crystal clear water, verdant mountains, billowing white clouds and transparently blue sky. Water that flows ceaselessly out of the Qilian Mountains makes up a non-exhaustible force that sustains this vigorous rangeland.

漫山紫花·
西藏芒康
Mountain cloaked in purple flowers — Mangkang, Tibet

这是由四川进入西藏时看到的景观，高山上漫山遍野的千里香杜鹃等杜鹃花披着紫色，非常奇特且漂亮。需要说明的是，杜鹃是草原上分布面积比例较大的灌木植被，在地类划分时，以覆盖率40%为界，大于40%的为灌木林地，不到40%的为高寒草原。

This amazing and impressive view was captured as I was about to enter Tibet from Sichuan, where the rolling mountains are almost completely covered under a purple blanket of *Rhodoendron thymifolium* and other rhododendrons blossoms. It is worth noting that rhododendron belongs to a shrubbery species that covers a comparatively large proportion of land here. Forty percent in shrubbery coverage is normally taken, in the categorization of lands, as the threshold for distinguishing shrubbery forests and alpine rangelands, with those higher than this percentage belonging to the former, and lower to the latter.

清晨炊烟·
青海班玛
Rising cooking smoke in early morning — Banma, Qinghai

沉睡的草原刚刚被第一抹阳光照亮，牛羊群、毡房被一个个清晰地显现出来。其实，牧人已经早起，缕缕炊烟下面，毡房里的酥油茶已经热气腾腾地端上来，牧区美好的一天就要开始了。

As the first beam of sunlight strikes on the sleepy rangeland, flocks of livestock and the tents of the herdsmen gradually reveal themselves. The pastoralists are already up and around. Beneath the billowing smoke that rises from the kitchens, breakfast consisting of hot ghee tea and other tasty food are already served. A brand-new and beautiful day is about to start on the pasture.

正在吃草的柯氏鼠兔·青海玉树

Ochotona koslowi foraging — Yushu, Qinghai

鼠兔是草原上司空见惯的啮齿类动物，而柯氏鼠兔的分布范围却极为有限，在中国主要分布在昆仑山一带，栖息于海拔4200米以上的高寒草原。柯氏鼠兔是现有的20余种鼠兔中最古老的遗留种。

Pika is a rodent that exist almost ubiquitous on rangelands, but *Ochotona koslowi* is found in very limited spaces. In China, they mainly inhabit in alpine rangelands that are located at 4,200m above sea level around the Kunlun Mountains. *Ochotona koslowi* is the oldest species among the approximately 20 pika species that still survive.

▼ 高山秃鹫•四川康定

Himalayan griffon —
Kangding, Sichuan

位于四川、西藏交界的折多山上，死去的牦牛尸体是高山秃鹫、兀鹫等食腐猛禽最好的食物。图为饱餐了一顿的高山秃鹫起飞的瞬间。

At the Zheduo Mountains that stand at the border between Tibet and Sichuan, dead bodies of yaks are the favorite food for raptors like Himalayan griffon (*Gyps himalayensis*) and Eurasian griffon (*Gyps fulvus*). Shown in the photo is a Himalayan griffon that is about to take on its wings after having filled its belly full with this delicious food.

▲ 红色草原•青海海晏

Rangelands of China's revolutionary tradition —
Haiyan, Qinghai

位于青海湖东北方向的大草原上牧草肥美牛羊壮，人们以金银遍地来形容这片美丽而富饶的土地，故得名"金银滩"。金银滩草原是名曲《在那遥远的地方》的诞生地，这里更孕育了新中国第一颗原子弹、氢弹，是中国第一个核武器研制基地。缘此，青海金银滩草原还被列入中国第一批"红色草原"名单中。图中草原远方高大烟囱处就是曾经的核实验基地之一。

The vast and fertile rangelands that lie to the northeast of the Qinghai Lake is teeming with lush grasses, strong and robust livestock, for which reason people name it as Jinyintan (a plain of gold and silver). Besides being the place to which the vintage song entitled *Over the Land in that Remote Corner* is dedicated, the Jinyintan was also the site of China's first nuclear weapons development base on which the country's earliest atomic and hydrogen bombs were developed. This is also the reason for which the Jinyintan Rangeland in Qinghai was listed among China's first batch of rangelands of revolutionary tradition. The towering chimney in the distance in the photo is just one of the earliest nuclear experiment bases.

高山草场·青海班玛
Alpine pasturelands —
Banma, Qinghai

这里位于青藏高原的东部、大渡河上游，尽管已经到了巴颜喀拉山的东端，河谷中也已经出现了森林，但在高山上占主要面积的还是高寒草原。由于降水充沛，草长势良好，这里是非常优良的高山草场。

This area is situated at the eastern QTP and the upper reaches of the Dadu River. Though already at the eastern end of the Bayan Har Mountains and forests are already appearing at the river valleys, the predominant view at the high mountains is still that of alpine rangelands. Abundant precipitation creates a very favorable environment for grasses to thrive, hence giving rise to some very good alpine pasturelands.

争雄·青海刚察
A showdown —
Gangcha, Qinghai

普氏原羚主要生活在青海湖畔的高寒草原和高寒荒漠草原地带。以禾本科、莎草科及其他沙生植物的嫩枝、茎叶等为食，冬季则以草原上的草茎和枯叶为食。普氏原羚发情交配期是12月至翌年1月，图为两只雄性普氏原羚为争取交配权在打斗。

The Przewalski gazelle(*Procapra przewalskii*), which chiefly inhabits the alpine rangelands and alpine desert rangelands surrounding the Qinghai Lake, lives on tender twigs and leaves of sandy-land-borne grasses belonging to the Gramineae and Cyperaceae families. In winter, they mainly feed on withered leaves and grass stalks available on rangelands. Mating season of this species falls between December and January. Shown in the photo are two male gazelles fighting for mating rights.

藏羚·青海可可西里
Tibetan antelope —
Hoh Xil, Qinghai

藏羚是青藏高原的精灵，栖息于海拔3700~5500米的高寒草原、高寒草甸和高寒荒漠草原地带，其通体被有厚密的绒毛，善奔跑，喜集群。其食物以禾本科植物和莎草科植物为主，冬季则啃食干草茎和枯叶。其忍耐干旱的能力较强，大多数时间是通过吃植物和吃雪获得水分。

Tibetan antelopes(*Pantholops hodgsonii*), fairies of the QTP, inhabit mainly on the alpine rangelands, alpine meadows and alpine desert rangelands lying 3,700–5,500 meters above the sea level. Their entire bodies covered with thick downy hairs, this gregarious species is particularly good at running. They feed normally on grasses belonging to the Gramineae and Cyperaceae families, but also on withered leaves and grass stalks available on rangelands in winter. Highly thirst-tolerant, they at most times obtain water through eating plants and snow.

3-5 高寒荒漠草原生态系统

ALPINE DESERT RANGELAND ECOSYSTEM

高寒荒漠草原生态系统是在高寒、干旱的大陆性高原气候条件下，由极其耐寒、耐旱的草本及小半灌木、灌木构成的植物群落。这里气候既寒冷又干旱，年平均气温－8~10摄氏度，月平均气温均在0摄氏度以下，全年没有无霜期，年降水量在50毫米以下，且全为固态降水。地下是多年冻土，土层薄，表土轻壤质并夹有小砾石，常含有盐分，只有那些极其耐寒、耐干旱的草本及小半灌木、灌木能够顽强地生长。在严酷的生态条件下，植物群落组成种类极少，生长稀疏，覆盖度一般10%左右，而在水分条件相对较好的地段，覆盖度可达25%。植物生长低矮，具耐寒抗旱的特性。

在青藏高原的高原面上，草原植被从东南向西北的递变规律为高寒草甸—高寒草原—高寒荒漠草原。再往西北，干旱加剧，植被进一步减少，海拔降低，演变为干旱荒漠草原生态系统和极干旱荒漠草原生态系统（见第二章），再进一步就进入荒漠生态系统（参见“一滴水生态摄影丛书”之一的《多样性的中国荒漠》，本书不再涉及）。

The alpine desert rangeland ecosystem refers to the plant community that exists under very cold and arid continental plateau climatic conditions, which is composed typically of extremely cold- and drought-resistant herbaceous plants, semi-shrubs and shrubs. The climate here is both cold and arid. With the average annual temperature ranging between -8 to 10 Celsius degree, and the average monthly temperature standing below 0 Celsius degree, there is hardly a single day that is frost-free throughout the year. The annual precipitation is lower than 50 millimeters, which mostly fall in solid forms. Underground is a thin layer of perm-frozen soil, where the gravel-mixed topsoil typically contains little loam. Because of the saline contents it often come with, only those extremely hardy, drought-resistant herbaceous plants, semi-shrubs and shrubs can manage to survive. Under such harsh ecological conditions, the plant community consists of a very small number of sparsely-growing species, normally just around 10% in coverage. In places where the moisture condition is relatively good, the coverage may occasionally reach 25%. The plants tend to be of low stalks, cold- and drought-resistant.

The surface of the QTP is covered from the southeast to the northwest by alpine meadows, alpine rangelands, and alpine desert rangelands consecutively. Moving further northwestwards, as the weather becomes even drier, plant coverage declines further, and elevation gets increasingly lower, desert rangelands will become the prevailing landscape (see Chapter 2), transferring ever closer to dry and extremely dry desert ecosystems (readers interested can refer to *Diverse Deserts of China* that has previously been published under the “A Single Drop of Water Ecological Photography Album series”).

高寒荒漠草原鸟瞰•青海都兰

A bird's eye view of alpine desert rangelands — Dulan, Qinghai

在青藏高原面上，随着降水量的减少，地面越来越趋于干旱，草原植被从东南向西北分别呈现的依次为高寒草甸—高寒草原—高寒荒漠草原。图为往西北方向即将进入柴达木盆地时的草原景观，草植被稀少，荒沙土露头。这个现象在提醒我们，荒漠离这里已经不远了。

As precipitation declines and the ground gets increasingly drier on the surface of the QTP, alpine meadows, alpine rangelands and alpine desert rangelands will appear consecutively in the southeast to northwest direction. Shown in the photo is the landscape that can be seen as one moves northwestwards and get closer to the Qaidam Basin, where the barren land is only sparsely covered by vegetation. What we see here reminds us that we are not far away from deserts.

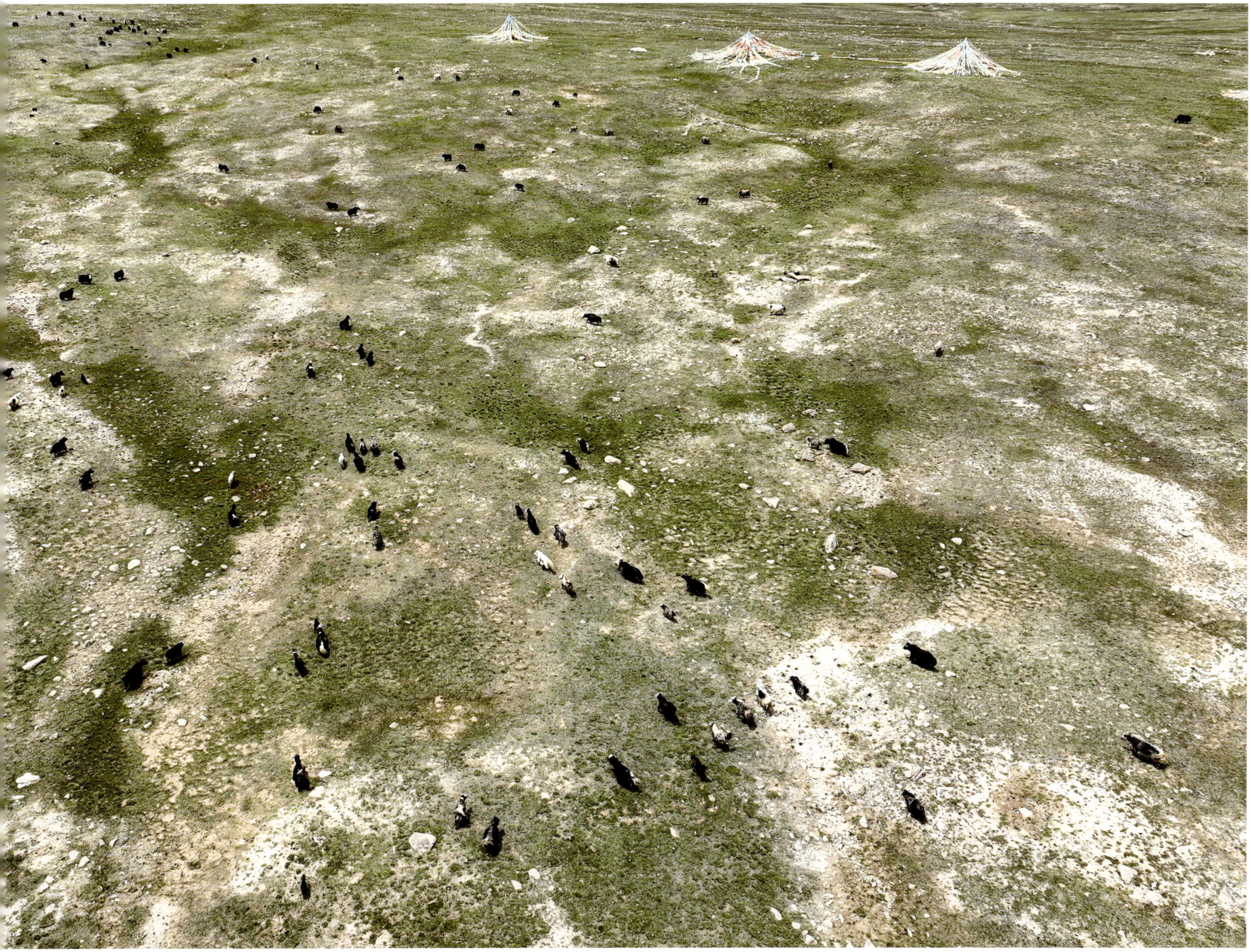

高山草场•甘肃肃南

Alpine pastureland —
Sunan, Gansu

比起祁连山西端南坡的森林草地，祁连山下部降水稀少、蒸发量大，处处呈现的是一片干旱荒漠草原的景观。

In sharp contrast to the forest grasslands situated at the southern slope of the west end of Qilian Mountains, the lower parts of the mountains tend to have very scarce rainfalls and high evaporation rates, which gives rise to the view of a landscape predominated by dry desert rangelands.

藏野驴母子群•青海格尔木

A family of Tibetan wild asses —
Golmud, Qinghai

藏野驴与蒙古野驴最直观的区别就在于其毛色更深，藏野驴栖息于海拔4200~5100米的高寒荒漠草原和山地荒漠带，比蒙古野驴的栖息地的海拔要高很多。藏野驴喜欢集群活动，通常出没于空旷的山间盆地、平缓的河谷阶地、丘陵、湖滩草地等。

Tibetan wild asses, which differ most visually from Mongolian wild asses in that their hairs tend to be comparatively darker in color, typically inhabit alpine desert rangelands and mountainous deserts that lie 4,200–5,100m above sea level, much higher than those inhabited by their Mongolian counterparts. Tibetan wild asses are fond of living in groups and often roam over open basins tucked between high mountains, gentle river valleys, hilly areas and beaches around lakes.

藏原羚·青海可可西里

Tibetan gazelles — Hoh Xil, Qinghai

藏原羚是典型的高寒荒漠草原动物，栖息于海拔3000~5750米的高山草甸、高寒草原及荒漠草原地带，主要以莎草科植物和禾本科植物等草类为食，体形比普氏原羚瘦小。

The Tibetan gazelle(*Procapra Picticaudata*), a typical alpine desert rangeland species that is slightly smaller than the Przewalski gazelle in size, chiefly inhabit alpine meadows, alpine rangelands and desert rangelands that lie around 3,000–5,750m above sea level. They mainly feed on grasses belonging to the Gramineae and Cyperaceae families.

旷野茫茫·
青海格尔木
Boundless wilderness — Golmud, Qinghai

大片延绵起伏的流动沙丘下面平洼之地是茫茫的高寒荒漠草原，藏野驴和藏羚在上面自由地生活着、跑动着，这里就是它们世代生存繁衍的美满家园。

Lying low below the vast stretch of rolling sand dunes is a boundless cold desert rangeland, over which countless Tibetan wild asses and Tibetan antelopes roam carefreely. This is the type of cozy and beautiful home in which they live and thrive.

▲ 高原兔·青海刚察

Lepus oiostolus — Gangcha, Qinghai

高原兔为青藏高原地区的特有种之一，栖息于海拔2100~4000米的高寒荒漠草原、高寒草甸草原、河谷及河漫滩灌丛中，隐蔽条件要求较高，故在芨芨草丛、灌丛生境中数量较多。

The *Lepus oiostolus*, a unique species indigenous to the QTP, mainly inhabits alpine desert rangelands, alpine meadow rangelands and river valleys that lie 2,100–4,000 meters above sea level. Being highly demanding of hiding places, they often found shelter among dense groves of *Neotrinia splendens* and other shrubberies.

◀ 羽毛三芒草·青海茫崖

Aristida pennata — Mangya, Qinghai

羽毛三芒草是沙漠里的先锋植物，生长于干旱、少雨、多风、基质松散、易于流动的沙丘和沙草地。为了保持水分的供需平衡，减少水分的蒸腾，它的叶片强烈纵卷，在植株基部形成不同高度的风积沙堆而起到固沙作用。羽毛三芒草在抽穗前为牲畜所喜食之牧草，马最喜食，其次为骆驼。

Aristida pennata is a pioneering species that grows well at dry, rain-scarce, windy places consisting of loose-binding and movable sandy hills and sandy grasslands. To cut down on evaporation and keep the supply and demands of water in balance, the leaves of this plant tend to curl notably. Wind-borne sands often settle down at the basal parts of the plant to form piles of varied height, hence playing a role to fix sands. This plant species, in the pre-sprouting phase, makes for tasty fodder for livestock. The animal that loves it best are horses, which is followed by camels.

3-6 森林草地生态系统

FOREST GRASSLAND ECOSYSTEM

在中国第一阶梯的青藏高原边缘，在向第二阶梯过渡的很多沟壑、峡谷中，由于海拔的逐步降低，东南季风带来的暖湿气流在爬升过程中产生降水，很多沟谷中产生了不少森林草地。如澜沧江上游的昂赛峡谷，长江重要支流——大渡河源流之一的马可河两岸，雅鲁藏布江中游部分沟谷，喜马拉雅山南坡的吉隆沟、樟木沟、亚东沟等，就分布着很多森林草地，虽然占比不大，却是一类难得的不可忽略的生态系统。这些森林草地主要是高寒草原向低海拔方向发展的结果，与内蒙古高原的森林草原是典型草原向湿热度增加方向（并非低海拔）发展的结果，原因并不相同。

At the numerous valleys and gorges situated at the transitional area between the edge of the first-terraced QTP and the second terrace, declining altitude and rising warm and humid airflow from the south-east monsoon bring about rich precipitation, which in turn creates highly favorable conditions for forest grasslands to appear in many of the valleys and gorges. Take for instance, the Angsai gorge at the upper reaches of the Lancang River, both banks of the Make River — a key tributary of the Dadu River that in turn feeds to the Yangtze River, some valleys at the middle reaches of the Yarlung Zangbo River, and such valleys as the Jilong, the Zhangmugou and the Yadong on the southern slope of the Himalayas, all of which are places where some forest grasslands can be found to exist. Though the proportion of such rangelands is not very large, they make up ecological systems whose values cannot be ignored. The presence of these forest grasslands is the result caused when alpine rangelands extend towards low-altitude areas, which is very different from what happens at the Inner Mongolian plateau, where the presence of forest rangelands is caused when typical rangelands extend towards areas that have more favorable water and heat conditions (rather than the altitude).

海拔分布最上限的柏树草地•青海杂多

The highest cypress grasslands — Zaduo, Qinghai

在三江源国家公园澜沧江源区的昂赛大峡谷的草原中，零散分布着大果圆柏、密枝圆柏等柏树种类的森林，其中，大果圆柏为优势树种，分布最广。这里是世界海拔最上限的柏树森林，这类森林和草地一起组成了特殊的柏树森林草地景观。

Over the rangelands located in the Angsai Valley at the source of Lancangjiang River — a part of the Sanjiangyuan National Park, there exists a huge patch of cypress forest that is composed of *Juniperus tibetica* and *Juniperus convallium*, the dominant and mostly-widely distributed trees being the former. This is the highest-lying cypress stand in the world, which combines with the grasslands to form this unique view of cypress grasslands.

无斑山溪鲵•
青海班玛
Batrachuperus karlschmidti —
Banma, Qinghai

在本区森林草地的小溪中，还生活着一种稀有的小型“娃娃鱼”——无斑山溪鲵，此种鲵全长仅151～220毫米，生活于海拔1800～4000米的山地小溪中，属两栖动物。

Living in the creeks that meander across the forest grasslands here is a precious small "baby-shaped fish" — *Batrachuperus karlschmidti*. This 151–220mm-long miniature amphibian mainly inhabits mountainous streams that lie 1,800–4,000 meters above sea level.

祁连山高山草地•
青海祁连
Alpine pasturelands at Qilian Mountains —
Qilian, Qinghai

祁连山地处青藏、蒙新、黄土三大高原交汇地带，是我国重要的生态功能区、水源涵养地和生物多样性保护优先区域。祁连山连绵近千千米，中国西北的几条大河都是从这里发源的，因此这里被称为“西北水塔”。由于坡位、坡向的不同，山上以青海云杉、祁连圆柏为主要优势树种的森林与大片的高山草地相间分布，极具特色。

The Qilian Mountains, located at the intersecting point of the Qinghai-Tibet, the Mongolia-Xinjiang and the Loess plateaus, is a region that provides crucial ecological services — water source conservation, biodiversity conservation and *etc.* — and is therefore regarded as an area that deserves priority in protection. Extending for thousands kilometers and being the source of the major rivers in north-western China, the Qilian Mountains is also hailed as the "Water Tower of the Northwest". Due to the influences of highly varied positions and directions of the mountain slopes, forests here, which are pre-dominated by Qinghai spruces, Qilian cypresses and other tall trees, would occasionally be intersected by vast stretches of alpine rangelands, hence presenting very unique and impressive views.

红隼•
西藏日喀则
Falco tinnunculus — Shigatse, Tibet

红隼体形矫健，飞行迅捷，广泛分布于山地森林、低山丘陵、草原、旷野灌丛草地、河谷和农田等地。红隼经常在空中盘旋，搜寻地面上的老鼠、小型鸟类、蛙、蜥蜴等动物，以及草地里的蝗虫、蚱蜢、蟋蟀等昆虫。

Falco tinnunculus, a beautifully-shaped and swift-flying raptor, is widely seen in places like mountainous forests, low hills, rangelands, shrubbery grasslands, river valleys and croplands. They tend to hover in the air and prey on small animals like rats, birds, frogs and lizards as well as on grassland insects like locusts, grasshoppers, crickets and *etc*..

人间天堂•
西藏林芝
Paradise in the secular world — Nyingchi, Tibet

林芝鲁朗被称为"神仙居住的地方"，两侧山地上是主要由灌木丛和茂密的云杉组成的"鲁朗林海"。宽阔的河谷中间是大片整齐划一的草甸，犹如人工修整过。草甸中，溪流蜿蜒，泉水潺潺，草坪上报春花、紫苑花、马先蒿花等多种野花盛开，构成一片由森林草地组成的"人间天堂"。

Lulang in Nyingchi is credited as "the abodes of the immortals", which is flanked on both sides by the "Lulang Forest Sea" composed of dense shrubberies and verdant spruce stands. Lying in the middle of the wide river valley is a vast stretch of neat and orderly meadow that looks as if well-attended by human beings. In addition to the meandering creeks and gurgling springs, the meadow is beautifully dotted with a large variety of wild blossoms like primroses, asters, and artemisia, hence often hailed as "the paradise in the secular world".

祁连圆柏·青海祁连

Juniperus przewalskii —
Qilian, Qinghai

祁连山的森林草地一般分布在海拔2800~3200米的山地，主要在阴坡、半阴坡比较湿润的坡面，主要树种为青海云杉。阳坡较为干燥，森林主要树种为祁连圆柏，草原占比较大，形成了特殊的祁连山森林草原复合生态系统。

Forest grasslands in the Qilian Mountains are typically found on the relatively humid mountain slopes of 2,800–3,200m-high mountains where the dominating tree species is *Picea crassifolia*. The sunny slopes are comparatively drier, where *Juniperus przewalskii* forms the dominant tree species. Grasslands tend to take up a very large proportion of the vegetation coverage, thus bringing into existence a very special type of compound ecosystem — the Qilian mountain forest rangeland ecosystem.

西藏马鹿·西藏桑日

Cervus elaphus wallichii —
Sangri, Tibet

西藏马鹿为极小种群野生动物，仅分布于西藏桑日的狭小地域，生活在高寒灌丛草地中。图中植被为金露梅灌丛，此地块的灌木覆盖度已经大于40%。因此，这也是森林草地生态系统的一类特殊表现形式。

Cervus elaphus, wildlife species of extremely small population, is found only within a narrow niche of alpine shrubbery grasslands in Sangri, Tibet. Shown in the photo is a grove of *Dasiphora fruticose*, where the coverage of shrubberies goes higher than 40%. In this sense, this also makes for a very special representation of the forest grassland ecosystem.

马麝·青海达日
Alpine musk deer —
Dari, Qinghai

马麝属于独栖动物，栖息于海拔3000~4000米与针叶林镶嵌的草甸及草原灌丛、裸岩等地，性胆小且孤独，多晨昏活动，善于奔跑，食性广泛，主要食用菊科、蔷薇科植物的嫩枝叶及地衣、苔藓等，是麝属中体形较大的一种。

Alpine musk deer(*Moschus chrysogaster*), a solitary-living species, typically inhabit coniferous-stand-intersected meadows, shrubbery rangelands and bare rocks that lie 3,000–4,000 meters above sea level. This timid, lone, agile and comparatively large-sized animal tends to be active at early morning and dusks, feeding broadly on the tender twigs and leaves of plants of the Asteraceae and Rosaceae families as well as on lichens, mosses and *etc.*.

森林草地牧民家·
青海杂多
Home of a pastoralist at forest grasslands — Zaduo, Qinghai

上游段的澜沧江蜿蜒流淌，陪伴着昂赛大峡谷一路往东南方向奔往四川盆地，峡谷两岸富饶的森林草地滋养着这里世世代代的藏族牧民。此时，毡房上升起了炊烟袅袅，好一幅世外桃源的真实景观，让人陶醉。

The upper reaches of the Lancangjiang River rushes by along the Angsai Grand Valley and makes its way southeastwards to the Sichuan Basin, leaving behind on both its sides vast stretches of fertile forest rangelands that have nourished generation after generation of Tibetan herdsmen. Light kitchen smoke is already rising from the tents, presenting a celestially realistic and intoxicating view of life on pasturelands.

小叶金露梅•
西藏工布江达
Dasiphora parvifolia —
Kongpo Gyamda, Tibet

小叶金露梅为落叶灌木，生于海拔3600～4800米的高山灌丛、高山草甸及山坡、路旁等处。其特点是极其耐寒，可以忍受-50摄氏度的低温。在青藏高原，生长有成片的金露梅，藏民们称之为“格桑花”。这类灌木的覆盖度在很多地方超过40%，被称为灌木林。

Dasiphora parvifolia is adeciduous shrub that grows at alpine scrubs, alpine meadows, mountain slopes and roadsides at 3,600–4,800 meters above sea level. This extremely cold-resistant species is able to survive at minus 50 Celsius degree. Vast patches of this plant(known among the Tibetan folks as *Gesanghua*) are often present on the QTP. As its coverage in many places exceeds 40%, it is often categorized as shrubbery forests.

银叶杜鹃和峨眉冷杉草地•
四川甘孜
Grassland dotted by silver-leaf rhododendron and Emei fir —
Ganzi, Sichuan

川西北为青藏高原的东南缘，海拔下降不少，受东南季风带来的暖湿气候影响，这里是该区自然温湿条件最好的地区，在高海拔的山顶、山坡草地上，银叶杜鹃和峨眉冷杉生长良好。其中，峨眉冷杉为中国四川特有种，而银叶杜鹃在西藏、四川、贵州、云南都有分布。

Situated at the south-eastern edge of the Qinghai-Tibetan Plateau, northwestern Sichuan Province is comparatively much lower in altitude and more highly subjected to the influences of warm and humid climate brought by southeastern monsoon, making it the most-favorable place in this region in terms of natural water and thermal conditions. The silver-leaf rhododendron(*Rhododendron argyrophyllum*) and the Emei fir(*Abies fabri*) thrive in high-reaching mountain tops and grassy slopes. Whereas Emei fir is an indigenous species unique to Sichuan Province, silver-leaf rhododendron can be found in a much wider area that extends across Tibet, Sichuan, Guizhou and Yunnan provinces.

3-7 可持续发展

SUSTAINABLE DEVELOPMENT

高寒草地生态系统主要分布在青藏高原及它的高山带，是青藏高原覆盖面积最大的生态系统。这里的地形条件复杂，气候条件严酷，各类生态系统的种类组成、结构及功能都非常特殊，保持水土、涵养水源、防风固沙、固碳生氧、保护高寒生物多样性等生态功能既特殊又丝毫不可削弱。这个"亚洲水塔""全球生态高地"对于保障中国国土安全乃至整个亚洲的生态安全有牵一发而动全身的极为重要的地位，是无可替代的。要认真贯彻执行《青藏高原生态保护法》，这对加强青藏高原生态保护，建设国家生态文明高地，促进经济社会可持续发展，实现人与自然和谐共生，都具有极其重要的意义。

本区基本属于中国自然动物地理区划中的古北界西部的青藏区，这里平均海拔达4000米以上，有"世界屋脊"之称，物种以耐高寒的种类为主。这里有藏羚、普氏原羚、野牦牛、藏野驴等食草动物，棕熊等杂食性大型动物，鼠兔、喜马拉雅旱獭等啮齿类动物，高山兀鹫、绿尾虹雉等高山留鸟，黑颈鹤、灰鹤、斑头雁等迁徙鸟类，它们都在这里栖息繁殖。这里还有绿绒蒿、塔黄、高山嵩草、垫状点地梅、雪莲、薹草、龙胆等高寒地带植物分布。

青藏高原是全球生态高地，草原是其绿色覆盖最广的主体生态系统，对中国、亚洲乃至全球的影响都是巨大的。因此，在自然保护、资源综合利用、环境整治及其可持续发展战略决策方面，应采取一系列最严格的保护策略。因此，有必要针对不同草原生态系统进行如下论述。

1. 流石滩稀疏植被生态系统的可持续发展对策

高山流石滩稀疏植被是一种特有的植被类型，对固定土壤和水土保持具有极其重要的作用。在目前全球气候变暖过程中，许多山地冰川处在退缩阶段，相应的流石滩面积就会不断地扩大，也会缓慢向垫状植被生态系统良性转化。所以，应该采取最严格的保护措施，禁止对于绿绒蒿、塔黄、雪莲等珍稀植物资源的采集利用，使流石滩稀疏植被在固定土壤和水土保持方面发挥更大的生态功能。

2. 垫状植被生态系统的可持续发展对策

由于垫状体紧贴地表和地下根系，具有极强的抗风蚀和保持水土的生态功能，同时垫状植被也属于先锋群落，在其生长过程中，能够不断地改善土壤条件，为其他植物定居创造优越条件，随着全球气候变暖，垫状植被生态系统也会缓慢向高寒草甸、高寒草原生态系统方面良性转化。因此，高寒垫状植被也应该被列入最严格保护的范围，严禁破坏和采挖利用。

3. 高寒草甸和高寒草原生态系统的可持续发展对策

高寒草甸和高寒草原的植被分布地域辽阔，占青藏高原总面积的一半以上，其生物生产力是所有高寒草地生态系统中最高的，可在地表15厘米深处形成紧密坚实而富有弹性的草结皮层，耐放牧践踏，具有涵养水源、保持水土、净化空气、保护高寒地区生物多样性的重要的生态功能。其上牧草营养丰富，具有高脂肪、高蛋白、高热值和低纤维素的特点，是高寒地区草地畜牧业的物质基础，也是当地老百姓生活可持续的根本保证，一定要在保护好自然生态环境不受破坏的前提下，稳定和减少畜牧业的规模，尤其要注重社会经济发展的正确导向。

4. 森林草地生态系统的可持续发展对策

本区的森林草地生态系统占比较小，但是特殊自然地理

草地上的岩羊与喇嘛• 青海兴海

Blue sheep and Lama monks on the pastureland — Xinghai, Qinghai

草原是"草—畜—人"和谐共生的生命共同体，在青藏高原这块特殊的生态高地上，人与自然关系的好坏不仅影响高原自身的生态环境和草地畜牧业的可持续发展，而且还影响中下游地区的社会安定和经济发展。因此，青藏高原具有中国生态安全和最重要保障的至高无上的地位，人与自然和谐共生尤为重要。图为喇嘛与岩羊群和谐相处。

Rangeland is a harmonious symbiosis composed of the mutually-dependent grasses, livestock and people. On this very special ecological high ground of the QTP, good people–nature relationship is not only decisive to the ecological wellbeing of the plateau and the sustainable development of the animal husbandry thereof, but is also of great relevance to the social stability and economic development of the lower and middle reaches of major rivers originated from there. For this reason, the QTP is of paramount importance to China's ecological security and sustainability, making it all the more imperative to keep the people–nature relationship at its most ideal status. The photo here is quite telling about the harmonious coexistence of blue sheep(*Pseudois nayaur*) and Lama monks.

条件下的产物，尤为珍贵。应严禁采伐森林，严格限制牲畜的数量。

青藏高原是中国黄河、长江、澜沧江及雅鲁藏布江等大江大河的发源地，被称为“亚洲水塔”，占青藏高原主要面积的草原生态系统在水源涵养方面的功能是异常重要、不可替代的。江河源区生态环境的好坏不仅影响高原自身的生态环境和草地畜牧业的可持续发展，还影响中下游地区的社会安定和经济发展，具有中国生态安全和最重要保障的至高无上的地位。

近几十年来，随着全球气候变化及人口激增，过度放牧，草地负荷超载，加之人类活动（开垦农田、采矿、滥挖药材等），已引起草地的严重退化。这不仅直接影响了草地畜牧业的可持续发展，而且导致生态功能失调、涵养水源能力下降、水土流失严重，甚至黄河上游断流及许多湖泊萎缩和消失。因此，有必要重新评估该区草地畜牧业发展战略和目标，国家必须要给予生态补偿政策和资金方面的大力扶持，在保护生态环境为第一要务的背景下，着力加强高寒草地生态系统的保护、治理与建设，促进传统畜牧业向生态畜牧业的根本转变，以达到生态、社会和经济协调发展的最终目的。

保护好这块“全球生态高地”，保护好中国生态安全最高地的原生自然生态系统是本区的最重要目标。本区生态系统服务和生态功能的核心主要应放在原生生态系统的保护、水源涵养和生物多样性保护方面。在实施措施上，坚持生态优先的原则，控制人类活动强度和载畜量，建议将大于 50% 的生态空间和承载力划入严格保护的生态保护红线之内，还给自然生态系统及野生动植物。应该严格限制脆弱高寒草原特色生态旅游业，采取禁牧、休牧、减牧、轮牧及杜绝乱樵滥采野生植物等严格管理措施，同时要下大功夫加速退化草地的生态修复，适度发展草原光电、风电产业，减碳减排，促进青藏高原社会经济绿色健康发展。

在完善青藏高原草原类型自然保护区空间布局和建设的基础上，要建设好以三江源、若尔盖、羌塘、青海湖、昆仑山、祁连山、香格里拉、贡嘎山、高黎贡山、冈仁波齐、珠穆朗玛峰、雅鲁藏布大峡谷等草原类和涉草类国家公园为主体的青藏高原自然保护地体系。

Alpine grassland ecosystems, which are mainly distributed on the Qinghai-Tibetan Plateau and tall mountains existing there, make up the ecosystem that covers the largest proportion of lands on the Plateau. Due to the highly complex terrain, the harsh climate, and the very special characteristics of the various ecosystems in terms of their composition, structure and functions, the task of protecting the various ecological functions — prevention of soil/water erosion, water source conservation, wind prevention and sand fixation, carbon sequestration and oxygen production, and alpine biodiversity conservation — is indeed very special, for which any slightest negligence will be too heavy a price for us to afford. This region, which is hailed as “the water towerof Asia” and “global ecological highland”, plays a crucial, decisive and irreplaceable role in safeguarding China’s homeland security and even the ecological security of Asia as a whole. We must conscientiously implement the *Qinghai-Tibetan Plateau Ecological Protection Law*, which is of great significance to strengthening ecological protection on the QTP, building a national ecological civilization highland, promoting sustainable economic and social development, and realizing the harmonious coexistence of human beings and nature.

This region falls basically, in terms of China’s natural zoogeographical zoning, within the Qinghai-Tibet district that lies in the western part of the palearctic realm, where is hailed — because of its 4,000–meters-high average altitude — as the “Roof of the World”, chiefly inhabited by highly cold-resistant species. Inhabited here are herbivores like Tibetan antelope, Przewalski gazelles, wild yaks, and Tibetan wild ass; large omnivorous animals like brown bears; rodents like pikas and Himalayan otters; residential alpine birds like Himalayan griffons and Chinese Impeyan; as well as migratory birds like black-necked cranes, gray cranes and bar-headed geese. This region is also the home of a huge variety of alpine plants, including *Meconopsis* spp., *Rheum nobile*, *Carex parvula*, *Androsace tapete*, *Saussurea involucrata* *Carex* spp., *Gentiana scabra* and *etc.*.

The Qinghai-Tibetan Plateau is a global ecological highland

where rangelands make up the ecosystem that has the widest green coverage and projects significant impacts on China, Asia and even the whole world. Therefore, a series of the most stringent protection strategies should be adopted with regard to the efforts for nature conservation, comprehensive resource utilization, environmental remediation and its strategic decision-making for sustainable development. Therefore, it is necessary to address each type of the rangeland ecosystems separately, as listed in the following.

1. Suggestions for the sustainable development of sparsely-vegetated alpine scree ecosystem

The sparsely-vegetated alpine scree ecosystem, an endemic vegetation type, plays extremely important roles in soil fixation and water and soil conservation. As increasingly more mountainous glaciers recede with the progress of global warming, the corresponding areas of screes will also expand, with potentials to develop positively towards the ecosystem of cushion plants. Therefore, the strictest protection measures should be taken to prohibit the collection and utilization of rare plant resources — for instance *Meconopsis* spp., *Rheum nobile* and *Saussurea involucrata* — and in turn better tap into their potentials in eco-services like soil retention, soil and water conservation and *etc*.

2. Suggestions for the sustainable development of alpine cushion vegetation ecosystem

Cushion plants, due to the fact that they often cling closely to the ground and underground root system, tend to have strong ecological functions — anti-wind erosion, water and soil conservation and *etc*. In addition to this, they also belong to pioneering plant community that are capable of improving the soil conditions in the process of growing and therefore creating favorable conditions for other plants to settle down and flourish. As global warming progresses, the ecosystems of cushion plants are also likely to develop positively and transform themselves to those of alpine meadows and alpine rangelands. Therefore, this type of vegetation should also be put under the strictest protection, with all activities — digging and other forms of destruction — that may have detrimental impacts on its health development put under strict ban.

3. Suggestions for the sustainable development of alpine meadow and alpine rangeland ecosystems

Alpine meadows and alpine rangelands exist extensively on the Qinghai-Tibetan Plateau, covering over half of the land territory of the Plateau. Boasting the highest biological productivity among all alpine grassland ecosystems, they are capable of forming a 15cm-thick grass crust layer that is compact, solid, resilient and highly resistant against the impacts of grazing. They also provide important ecological functions like water conservancy, soil and water conservation, air purification, and protection of the alpine biodiversity. Fodders grown here are nutritious — with high contents of fat, protein and calories but low in fiber contents, laying down a solid materialistic foundation for local animal husbandry industry and sustainable livelihood for local people. Subject to the prerequisite that the natural ecological environment is not damaged, the scale of livestock husbandry must be kept steady and gradually cut down. In particular, attentions must be paid to make sure that the socio-economic development follows the right course.

4. Suggestions for the sustainable development of forest grassland ecosystems

The forest-grassland ecosystems of the region, which account for a relatively small proportion of the total, are the product of special natural geographical conditions and are particularly valuable. Deforestation should be strictly prohibited and the number of livestock should be strictly controlled.

The Qinghai-Tibetan Plateau, from where most of China's major rivers — the Yellow River, Yangtze River, Lancang River and YarlungZangbo River — have their sources, is credited as the "Water Tower of Asia". Rangeland ecosystems, which cover a vast majority parts of the Plateau, play an exceptionally important and irreplaceable role in water conservation. The quality of the ecological environment of the headwaters is decisive not only to the ecological environment of the plateau itself and the sustainable development of local animal husbandry, but also to the social stability and economic development of the middle and lower reaches of the rivers. Their important role in safeguarding the country's eco-security can never be overestimated.

Global climate change, population explosion, overgrazing and exploitative utilization, all of these have led to serious grassland degradation, which was made even worse by the irrational interference of other human activities (like cropland cultivation, mining, indiscriminate digging of medicinal herbs and *etc*.). This has not only put direct negative impacts on the long-term sustainability of local husbandry industry, but also resulted in ecological dysfunctions — shrinking water conservation capacity, ever serious water/soil erosion, drying out of tributaries that feed into the upper reaches of the Yellow River, and shrinking sizes and even total disappearance of numerous lakes. Therefore, it is necessary to reassess the development strategy and objectives of grassland animal husbandry in the region. In the context of prioritized ecological environment conservation, it is imperative for the State to work out more robust initiatives — such as ecological compensation policies and supportive funding — so as to strengthen the protection, management and construction of alpine grassland ecosystems, and to promote the fundamental transformation of traditional animal husbandry into eco-animal husbandry, hence contributing to the final achievement of our ultimate goal — coordinated ecological, social and economic development.

The overriding goal of this region is to keep the "global ecological highland" — together with all the native natural ecosystems that are decisive to China's eco-security — well protected. The core of ecosystem services and ecological functions of the region should mainly focus on the protection of native ecosystems, water conservation and biodiversity protection. With regard to its specific implementation, the ecological-priority principle must be consistently adhered to while controlling the intensity of human activities and limiting the carrying capacity of rangelands. It might be proposed that over 50% of the ecological space and carrying capacity should be identified as the exclusive lands of natural ecosystems and wildlife, where human access should be strictly banned. In addition to placing strict restrictions on eco-tourism development in fragile and vulnerable alpine rangelands, corresponding measures — such as grazing ban, grazing moratorium, grazing mitigation, rotation, logging ban, digging ban and *etc.* — must also be put in place and earnestly implemented. Meanwhile, efforts must also be strengthened to restore the degraded grasslands through ecological-based means by encouraging scientifically-planned development of renewable energies like solar- and wind-generated electricity, thus laying down a solid foundation for the green and healthy socio-economic development of the QTP.

Besides further improving and optimizing the spatial layout of rangeland-featuring nature reserves in the QTP region, efforts are also urgently needed to put in place a natural protected area system that centers around the QTP rangeland and rangeland related national parks, including the Sanjiangyuan, Ruoergai, Qiangtang, Qinghai Lake, Kunlun Mountains, Qilian Mountains, Shangri-La, Gongga Mountains, Gaoligong Mountains, Gonzaga Mountains, Mt. Everest, and the Great Canyon of the Yarlung Zangbo and *etc.*

和谐发展•青海海晏
Harmonious development — Haiyan, Qinghai

在保护生态环境为第一要务的背景下，应该着力加强高寒草地生态系统的保护、治理与建设，促进传统畜牧业向生态畜牧业的根本转变，促进青藏高原社会经济绿色健康和谐发展，因此，保护好中国生态安全最高地是本区的最重要目标。图中与列车相迎面的是普氏原羚。

Under the historical background when ecological environment conservation is regarded as the top priority, efforts must be strengthened to protect, manage and build the alpine grasslands ecosystem. Through encouraging the traditional husbandry to transform to eco-husbandry, the society and economies of the QTP will undergo green, healthy and harmonious development. For this reason, safeguarding the ecological security of this highest-lying region of China should be taken as the ultimate goal of all of our work. Shown in the photo is a Przewalski gazelle, which looks as if running up to greet the train.

CHAPTER 4 第四章

东北华北平原山地丘陵草原区

The Northern and Northeastern Plain and Hilly Rangeland Region

阳光初照·黑龙江五常
Sunrise — Wuchang, Heilongjiang

凤凰山主峰海拔1692.6米，为张广才岭之首，山顶是一大片开阔的高山草甸。由于草本生长茂盛厚实，素有"高山地毯"之称。此时，东方的山脊上射出了清晨的第一道阳光，将凤凰山这片最高处的高山草甸染成了金黄色。

The main peak of Fenghuang Mountain, 1692.6 meters in altitude, is the tallest one among the Zhangguangcai Mountain Ranges, on top of which lies a vast stretch of open alpine meadow. The thriving and verdant vegetation coverage over the meadow has earned it a reputation as "the alpine carpet". Shown in the picture is the time when the first beam of sun ray rises over the ridge in the east, enshrouding this highest-lying alpine meadow over the Fenghuang Mountain in bright golden color.

II
I
北京★
Beijing
IV
III
V
南海诸岛
China South Sea Islands
1:64 000 000
图例 Legends
I 内蒙古高原草原区
The Inner Mongolian Plateau Rangeland Region
II 西北山地盆地草原区
The Mountainous and Basin Northwestern Rangeland Region
III 青藏高原草原区
The Qinghai-Tibetan Plateau Rangeland Region
IV 东北华北平原山地丘陵草原区
The Northern and Northeastern Plain and Hilly Rangeland Region
V 南方山地丘陵草原区
The Southern Hilly Rangeland Region
国界 National Border
1：32 000 000

本区西缘的北部为大兴安岭主脊，往南后沿科尔沁沙地东缘，接燕山、太行山至黄土高原南，走关帝山接乌鞘岭，西到贺兰山、陇东丘陵止，南到秦岭—淮河一线，包括这条线以东和以北的大部分东北、华北地区。其中，三江平原、松嫩平原、西辽河平原、华北平原、黄河下游平原都是河流冲积平原，其上普遍沉积很厚的第四纪松散沉积物，地形比较平坦，土壤有机质与土壤肥力都较高，水资源比较丰富。

在东北部的中温带半湿润气候区，除在局部隆起的山地丘陵如兴安林缘、长白山等上面分布着森林草地生态系统之外，还分布着暖性草丛生态系统和暖性灌草丛生态系统。

在华北部的暖温带半湿润气候区，暖性草丛生态系统和暖性灌草丛生态系统则主要分布在晋陕黄土高原、山东、辽东半岛低山丘陵及燕山、太行山、秦岭北坡等山地，土壤主要为褐土、棕壤，局部为黄棕壤。其地带性的植被为落叶阔叶林，但由于历史上长期的砍伐、采挖、开垦等原因，原始的地带性植被已所剩无几。在自然条件下，被破坏的天然植被恢复到原来的顶极群落在短期内已不可能，从而形成了次生的但相对稳定的暖性草丛与暖性灌草丛生态系统。

另外，本地区还零散分布典型草甸、沼泽草甸和盐生草甸等草甸生态系统类型。

The western border of this region starts in the north from the main ridge of the Daxinganling Mountain Ranges and extends southwards — through the east edge of the Horqin Sandy Land, the Yanshan Mountain, the Taihang Mountain, the southern part of the Loess Plateau, further through the Guandi Mountain, the Wushaoling Range, the Helan Mountain and the Longdong hilly areas — until it reaches the Qinling-Huaihe line. All places lying to the east and north of the previous-mentioned line, or most parts of North China and Northeast China, fall within this region. The Sanjiang Plain, the Songnen Plain, the Xiliaohe Plain, the North China Plain and the Plain of the Yellow River Lower Reaches are all comparatively flat alluvial plains that generally come with thick layers of Quaternary sediments, where the soil is typically rich in both organic matter content and fertility and has abundant water resources.

In the semi-humid mid-temperate climate zone in the north-eastern part of this region, forest grassland ecosystems are found in certain undulated hilly areas like the Xingan Forest Rim and the Changbai Mountain. Besides this, warm tussock ecosystems and warm shrubbery and tussock ecosystems are also widely present.

In the semi-humid warm-temperate climate zone in the North China part of this region, warm tussock ecosystems and warm shrubbery and tussock ecosystems mainly exist in the Shanxi-Shaanxi parts of the Loess Plateau, in the low hilly areas of the Shandong and Liaodong peninsulas, as well as in the mountainous areas of the Yanshan Mountain, the Taihang Mountain and the northern slope of the Qinling Mountain, where the predominant type of soils are cinnamon soil and brown loam, with yellow-brown loam occasionally also seeable in certain local areas. The zonal vegetation in this region is deciduous broad-leaved forests, a major part of which has unfortunately been lost over the long years due to excessive logging, digging and land reclamation activities. Given that it is hardly possible — under natural conditions — for natural vegetation that has been destroyed to recover in short terms to its primary plant communities, secondary but comparatively stable warm tussock ecosystems and warm shrubbery and tussock ecosystems have over the years emerged in their places.

Randomly scattered in this region is also the meadow ecosystem like that of typical meadows, marshy meadows and saline meadows.

4-1 森林草地生态系统

FOREST GRASSLAND ECOSYSTEM

本区的森林草地生态系统，主要分布于山地的阴坡与丘顶，同时也见于一些山地森林的上部和湿地的周边。

有别于内蒙古高原面上的草原生态系统，该区的草地面积较小和零散，海拔相对较低，降水量丰富，一般在有一定坡度的山地上分布。这里的森林草地生态系统是草原生态系统中最湿润、生态条件最好的一种类型，植物种类组成比较丰富。

森林草地带是最好的宜牧地，往往也被认为是宜农、宜林地。由于人口密集，人们的生活和生产活动产生大量的放牧、割草、开垦、砍伐等现象，导致这里土地退化，土地生产力急剧下降，甚至丧失。

The forest grassland ecosystem in this region is primarily found at the shady slopes as well as the top of mountainous areas, occasionally also observable at the upper boundaries of forests or at the perimeters of wetlands.

Unlike the rangeland ecosystem typically found on the Inner Mongolian Plateau, grasslands in this region tend to be relatively smaller in size, fragmented in geographical distribution, lower in altitude and abundant in precipitation — mostly existing on mountain slopes that come with a certain gradience. The forest rangeland ecosystem in this region belongs to the most humid type that is endowed with the most favorable ecological conditions, hence also comparatively more diversified in its plant composition.

Forest grasslands are unquestionably the best grazing lands and are sometimes also considered as favorable sites for agricultural and forestry development. High population density in this region — together with the extensive livestock grazing, fodder harvesting, land cultivation and logging that are often associated with people's livelihood and productive activities — has led to land degradation, as shown by the sharp decline, even total loss of land productivity.

森林草地·吉林蛟河
Forest grasslands — Jiaohe, Jilin

老爷岭为长白山支脉，这里属于牡丹江流域的汇水区范围，为著名的东北平原提供源源不断的水源。图为山顶平缓的山脉，山上的森林草地沐浴在晨光之中。

The Laoyeling Mountain, a branch of the Changbai Mountains, is situated within the catchment of the Mudanjiang River, providing stable water supply to the highly reputed Northeast Plain of the country. Shown in the photo is the gently-extending mountain at its top, where a vast stretch of forest grassland is bathed cozily in the morning sun.

"三元"牧场·黑龙江嫩江

The "Tri-element" pastureland — Nenjiang, Heilongjiang

这是林地、草场、农地"三元"混合的生态系统，这里的老百姓依靠这个"三元"生态系统维持着半农半牧的生活方式。这种生活方式在我国东北森林草原交错区域常见。

Shown in the photo is a mixed ecosystem composed simultaneously of triple elements — forests, pasturelands and croplands — that local folks depend on for their semi-agricultural and semi-husbandry-based livelihood. This sort of livelihood is very common in certain areas of Northeast China where forests and grasslands coexist side by side.

灵山秋·北京门头沟

Autumn at the Lingshan Mountain — Mentougou, Beijing

灵山顶峰海拔2303米，是北京的第一高峰，其海拔高度超过本区的泰山、华山等著名山峰。这里以海拔1900米以上的高山草甸最为出名，图为灵山主峰附近的森林草地生态系统。

The 2303m peak of the Lingshan Mountain is the highest in Beijing, higher in altitude than many widely-acclaimed mountains — Mount Tai and Mount Hua for instance — in this region. This place is especially well-known for the alpine meadows that lie over 1,900 meters above sea level. Captured in the photo is the forest grassland ecosystem that exists in the vicinity of the main peak.

驯鹿·内蒙古鄂温克旗

Reindeer —
Ewenke Banner, Inner Mongolia

这里属于中国草原地理区划中东北华北平原山地丘陵草原区的兴安林缘草原亚区。作为寒温带大型食草动物的驯鹿，一般栖息在寒温带针叶林中，处于半野生状态。驯鹿的食物主要是石蕊，也吃苔藓蕨类植物、蘑菇及嫩枝叶。鄂温克族是中国境内唯一饲养驯鹿的民族，驯鹿在其日常生活和生产中扮演着重要的角色。

This place belongs, in terms of China's rangeland geographical zoning system, to the sub-zone of the Xing an Forest Rim Rangeland that fall within the Northern and Northeastern Plain and Hilly Rangeland Regions. A large herbivore of the cold-temperate zone, reindeers(*Rangifer tarandus*) generally inhabit — in semi-wild manners — the coniferous forests in the cold-temperate zone, feeding mainly on *cladonia* spp., mosses, ferns, mushrooms and tender twigs and leaves. The Ewenki people are the sole ethnic group in China that raises reindeer, with the latter playing an important role in their daily life and production.

南草北木·
陕西柞水

Grass in the south and wood in the north — Zhashui, Shaanxi

在秦岭主脉附近的山头上，山脊一边是草，另一边是树，景观截然不同，这个现象被人们称为“南草北木”，这个“草”就是指草地，这个“木”就是指乔灌木。由于南坡迎风干旱，北坡避风湿润，造成的小气候类似的现象使山脊两边出现截然不同的植被类型，其实，在西北的很多大山，如天山、祁连山等，我们也能常常看到这种类似现象。

Views on the southern and northern slopes of the Qinling Mountain Ranges vary distinctively near the top — covered by grasses on one side and trees on the other, which is commonly known among people as the phenomenon of “southernly grasses and northernly wood”, with the “wood” referring both to tall trees and shrubberies. This is the result caused by the different micro-climate on the two sides of the mountain ridge — the wind-facing southern slope being dry and the wind-proof northern slope being humid. In fact, similar phenomenon can frequently be seen in many high mountains in northwestern China, like the Tianshan Mountains, the Qilian Mountains and *etc.*.

山花烂漫·吉林抚松

Charming wildflowers over the alpine meadows — Fusong, Jilin

长白山以五个植被带（落叶阔叶林带、针阔叶混交林带、针叶林带、岳桦林带和高山苔原带）的垂直植被分布而闻名天下。其实，长白山的森林草地也是一类难得的生态系统。图为山花烂漫的高山草地在周边森林的护卫下，显得更加美丽动人。

The Changbai Mountain is widely known across the world for its 5 vertically distributed vegetation belts (that of deciduous broad-leaved forests, mixed coniferous and broad-leaved forests, coniferous forests, *Betula ermanii* forests, and alpine tundra). As a matter of fact, the forest grasslands here is also a rare and precious ecosystem that is not often seen in other places. Shown in the photo is a wild flower-dotted alpine meadow, which looks all the more charming and graceful against the dense forests that stand toweringly around it.

草地梅花鹿•
黑龙江阿城
Sika deer grazing over the grasslands — Acheng, Heilongjiang

东北梅花鹿是东北地区较大的食草动物，生活于森林边缘和山地草原地区，白天多选择在向阳山坡上茅草深密与体色相似的地方栖息。公鹿能够生成鹿角，成熟时可达4~5个叉。

Sika deer, the large herbivore indigenous to northeastern China, typically live on the edges of forests or mountainous rangelands, where they would find shelters on sunny slopes amidst dense grass groves that are similar in color to their own. Bucks normally have antlers — with up to 4-5 forks at maturity — over their heads.

牛皮杜鹃•
吉林长白
Rhododendron aureum — Changbai, Jilin

牛皮杜鹃为常绿矮小灌木，喜凉爽湿润的气候，不怕霜冻，惧酷热干燥，生于海拔1000~2500米的高山草原地带或垫状植被带上。每到夏季，在长白山的高海拔处，牛皮杜鹃开得漫山遍野。

Rhododendron aureum, a frost-resistant and heat/drought-aversive evergreen small shrub that grows well in cool and humid climate, tend to thrive on alpine rangelands or cushion vegetation that lie 1,000–2,500 meters above sea level. The high-altitude places in the Changbai Mountain will be densely dotted with charming blossoms of this shrubbery species.

4-2 暖性草丛生态系统和暖性灌草丛生态系统

WARM TUSSOCK ECOSYSTEM AND WARM SHRUBBERY AND TUSSOCK ECOSYSTEM

在中温带和暖温带湿润、半湿润气候条件下，水、热条件比较优越，气候温暖而湿润，在降水丰富的条件下，以多年生草本植物为主形成的生态系统被称为暖性草丛生态系统。这类生态系统一般都是在此类气候条件下，森林生态系统在被长期破坏以后形成的较为稳定的次生生态系统。在暖性草本植物中散生灌木或零星乔木的被称为暖性灌草丛生态系统，没有和极少有灌木的则为暖性草丛生态系统。

暖性草丛生态系统和暖性灌草丛生态系统，主要分布于辽东半岛与辽西低山丘陵，山西恒山及管岑山、燕山和长城以南的河北、北京、河南、陕西、山东等省低山和中山山地。这些地区气候温暖，降水较丰，土壤主要为不同母质来源的褐土，土层浅薄不等，肥力中等，呈微碱性反应。

从分布地区的光、热、水条件看，暖性灌草丛生态系统相比暖性草丛生态系统的条件要优越。

Under the humid and semi-humid climates of the mid-temperate and warm temperate climate belts that are endowed with favorable water and heat conditions — as shown by abundant precipitations and balmy weather, the ecosystem composed primarily of perennial herbaceous plants is generally referred to as the warm tussock ecosystem. This type of ecosystem is usually a relatively stable secondary ecosystem that is formed after the primary forest ecosystem that used to exist under this climatic condition was gradually damaged over long periods of time. Whereas the ecosystem that has few or no shrubberies is referred to as the warm tussock ecosystem, the ecosystem that is randomly dotted with shrubs or sparse tall trees is called the warm shrubbery and tussock ecosystem.

These two types of ecosystems are mainly found to exist in the Liaodong peninsula, the Liaoxi low-lying hills, the Hengshan Mountain in Shanxi, as well as in the low and medium hilly areas in parts of Hebei, Beijing, Henan, Shaanxi, Shanxi and Shandong provinces that lie to the south of the Hengshan Mountain, the Guancen Mountain, the Yanshan Mountain and the Great Wall. These areas are generally warm in climate and comparatively abundant in precipitations. Soils found here mainly belong to variously-sourced cinnamon soils that differ more or less in thickness. Being medium in fertility, the soils are slightly skewed toward the direction of being alkaline.

Viewed from the combination of light, heat and water conditions, that of the warm shrubbery and tussock ecosystem is comparatively more favorable than that of the warm tussock ecosystem.

草坡上的马鹿• 黑龙江伊春
Red deer on mountain slope — Yichun, Heilongjiang

这里是中温带低海拔温润的小兴安岭，金黄色的旭阳照射着这片茂密的草丛。阴影中马鹿在回头张望，身体上部出现在阳光里，与笔者的镜头正好对上。

This photo was taken at the warm and humid low-lying part of the Little Xingan Mountain Range situated in the mid-temperate climate zone when the golden early morning sun cast its first rays on the dense growing grasses. The red deer in the shadow turned its head coincidently back towards the camera lens, with the upper part of its body exposed under the sunlight.

大鸨欲飞·辽宁阜新

A great bustard about to take on its wings — Fuxin, Liaoning

大鸨为大型地栖鸟类，主要栖息于开阔草原、稀树草原和荒漠草原，在冬季和迁徙季节也出现于河流、湖泊沿岸和邻近的草丛、湿草地，常成群活动，善于奔跑，主要吃植物的嫩叶芽、种子等植物性食物及昆虫、蛙等动物性食物。

The great bustard (*Otis tarda*), a large ground-dwelling wild bird, mainly inhabit open grasslands, savannas and desert rangelands, and may also appear at rivers, lakesides as well as within tussock during winter and migration seasons. This fowl species, gregarious and excelling in running, feeds mainly on herbaceous food like the tender leaves and fruits of plants as well as on insects and frogs.

太岳山暖性草丛·山西沁源
Warm tussocks at the Taiyue Mountain — Qinyuan, Shanxi

位于太岳山的花坡草原是国家公布的首批国家草原自然公园试点之一，由于这里是抗日战争时期太行地区通往延安的秘密交通线的重要段落，被列入中国第一批“红色草原”名单。

The Huapo Rrangeland, situated at the Taiyue Mountain, ranks among the country's first batch of pilot natural parks featuring rangelands. In addition, being a critical passage on the secret communication line leading from the Taihang Mountain to Yan'an during the war against Japanese aggression, this place is listed among the country's first batch of rangelands of revolutionary tradition.

西拉木伦河源区·辽宁朝阳
Source of the Xilamulun River — Chaoyang, Liaoning

西拉木伦河为西辽河北源，这里水草丰美。沿河两岸地势平坦，广泛分布着山丘草原和黄土丘陵地貌，暖性草丛和森林草地为这里的主要植被类型。

The Xilamulun River, the northern source of the Xiliaohe River, is a place of abundant water and flourishing grasses. Extensively distributed over the flat terrain on both sides of the river are vast stretches of hilly rangelands and loess landforms, over which the dominant vegetation types are warm tussocks and forest grasslands.

▲ 奔跑•吉林延边
On the flight — Yanbian, Jilin

在灌草丛中，狍子时而急速奔跑，时而高高跃起。这是因为草丛中分布着不少高大灌木丛，促使狍子不得不这样做。狍子性情胆小，日间多栖于密林中，早晚时分才会在空旷的草场或灌木丛中活动。

Shown in the photo is the roe deer(*Capreolus pygargus*) that leaps and bounces among shrubbery tussock grasses — it has to do so because of the presence of many tall shrubberies over the grasslands. Being a timid species by nature, the roe deer tends to hide among dense forests during daytime and come out only in early morning or at dusk to forage on open grasslands or among shrubberies.

◀ 黄土高原农牧交错带•甘肃平凉
The agricultural-husbandry borderline on the Loess Plateau — Pingliang, Gansu

这里是半湿润半干旱区黄土高原农牧交错带的景观，是"胡焕庸线"经过的地方。山坡草地与开垦的农地相间，沟谷里较阴湿的地方有树木生长，这里也是水土流失比较严重的地区。近些年来，通过实施退耕还林还草的政策，这里生态环境得到了大大地改善。

Shown in the photo is the view of a place on the semi-moist and semi-arid Loess Plateau where agriculture-based culture is gradually being replaced by a husbandry-based one, as marked by what's known as the Hu Huanyong dividing line. Grassy mountain slopes intersect with arable farmlands. Although there are also trees in low-lying shady and humid valleys, these places are also highly susceptible to water and soil erosion. Thanks to the implementation of croplands-to-forests/grasslands policy over recent years, the ecological conditions here have been notably improved.

长城脚下·河北滦平
At the foot of the Great Wall — Luanping, Hebei

图为太阳光柱下的长城南坡，这里的主要植被是暖性灌草丛，尽管冬季降过雪后的山坡一片枯黄，但是我们仍然能看到草丛中的灌木分布，在较阴湿的沟底、阴坡，甚至还有零星乔木生长。

Shown in the photo is the southern slope of the Great Wall bathed in strong and direct sunlight, where the dominant vegetation coverage is composed of warm shrubbery and tussock grasses. In spite of the bleak view of withered grasses on the snow-covered slope during winters, we can still obviously see the shrubberies strewn among the grasses. At the shady and humid bottom of the valleys and over the shady slopes, sparsely growing tall trees can also be seen.

4-3 草甸生态系统

MEADOW ECOSYSTEM

由于水分的逐步增多，使生态系统向典型草甸生态系统、沼泽草甸生态系统、沼泽湿地生态系统方向变化。沼泽湿地生态系统已属于湿地范畴了（参见“一滴水生态摄影丛书”之一的《多样性的中国湿地》，本书不再涉及）。

在沿海及其他一些地势低洼的内陆地区，还存在大面积的盐碱土或盐化草甸土，这类土壤含盐量较高，比较湿润，或有短期的季节性积水。在这种条件下，只有比较耐盐的或盐生、泌盐的植物才能正常生长，被称为盐生草甸生态系统。

在本区，主要分布典型草甸生态系统、沼泽草甸生态系统和盐生草甸生态系统。

Ever increasing water causes the zonal ecosystem to gradually shift towards that of typical meadows, marshy meadows and marshy wetlands, the last among which falling already within the category of wetlands (readers interested can refer to *Diverse Wetlands of China* that has previously been published under the “A Single Drop of Water Ecological Photography Album series”).

At the coastal and certain low-lying inland areas in this region, there also are vast stretches of saline or salinized meadow soils that typically come with relatively higher salt- and moisture-content and that tend to have short-term seasonal waterlogging. Only salt-tolerant, salt-borne and salt-secreting plants can survive such conditions, for which reason this type of ecosystems are known as that of saline meadows.

The ecosystems of typical meadows, marshy meadows and saline meadows are the major types of the meadow ecosystem normally found in this region.

红绿盐生草甸•辽宁盘锦

The red and green saline meadow — Panjin, Liaoning

辽河从上游带来的有机物与无机物在入海处沉积，加上潮汐海水的影响，形成了这片盐生草甸。碱蓬在每年的秋季都会将这片滩涂染红，其嫩芽和种子为丹顶鹤提供了最好的食物，这里也是黑嘴鸥的主要繁殖地。

The organic and inorganic materials that are flushed down by the Liaohe River from its upper reaches deposit at the river mouth, which — together with the impacts of tidal water — brings into shape this saline meadow. Each year when autumn sets in, the beach will be turned into a red carpet made of *Suaeda glauca*, the tender shoots and seeds of which are favorite food for red-crowned cranes. Besides, this beach is also an important breeding ground for another key protected bird species — black-billed gull.

多种鹤的集群•
山东东营
An assembly of various cranes — Dongying, Shandong

在黄河入海口的河滩草甸上，可以看到白鹤、白头鹤、灰鹤等多种鹤在这里集群。黄河三角洲是东亚—澳大利西亚鸟类迁徙路线上非常重要的驿站，每年都有大量的各种各样的迁徙鸟类途径这片广阔无垠的沼泽草甸。

Over the meadows lying at the mouth of the Yellow River, a huge variety of cranes — like whooping cranes, white-headed cranes, gray cranes and *etc.* — would often gather in assembly. Being a critical stop on the East Asian–Australasian migratory route, the boundless meadows nourished at the Yellow River Delta witness almost countless migratory birds — varying in species and shapes — to arrive and depart from here each year.

盐生草甸跑牛羊•
黑龙江齐齐哈尔
Herds of sheep and cattle over saline meadows — Qiqihar, Heilongjiang

深绿色的草甸与白花花的盐碱地相互交错、相互交融着，这是从航拍角度看到的一类盐生草甸的特殊景观。很难想象，这样贫瘠的盐渍化土地上，一群群牛羊照常能够找到它们需要的草食，不离不弃地生活在这片盐生草甸上。

Dark green meadows intersect with glaringly white salinized lands in irregular patterns to form a very unique view of saline meadows, which becomes all the more obvious and impressive when looked down from the air. It is hardly imaginable that, despite of the seemingly bleak and barren salinized land, herds and herds of cattle and sheep can still find abundant supply of fodders most palatable and satisfying to them.

雄性麋鹿群·江苏大丰
A herd of male elks —
Dafeng, Jiangsu

麋鹿为大型食草动物，喜群居，善游泳。重引入的麋鹿自由自在地生活在这片沼泽草甸上，至今，大丰拥有的麋鹿数量已经占到了世界总数的1/3，被授予"中国麋鹿之乡"的光荣称号。2019年，苏北这片包括大丰保护区、盐城保护区、条子泥保护区、东台高泥湿地及东台条子泥湿地在内的滩涂，以中国黄（渤）海候鸟栖息地（第一期）的总称，被列入《世界遗产名录》。

Elks(*Elaphurus davidianus*) are the large gregarious herbivore that excels in swimming. Since their successful reintroduction, elks have been living carefree on the vast marshy meadows in Dafeng, a place hailed as the "land of elks in China" for its large elk population — accounting up to 1/3 of the world's total. In 2019, this extensive mudflat in northern Jiangsu province — including the Dafeng Reserve, the Yancheng Reserve, the Tiaozini Reserve, the Dongtai Gaoni Wetland and the Dongtai Tiaozini Wetland — was listed among the world heritages, entitled in general as China Yellow (Bohai) Sea Habitat of Migratory Birds (Phase I).

大片的塔头地·黑龙江扎兰屯

Vast areas of "Pagoda-head land" — Zhalantun, Heilongjiang

"塔头"是一种高出地面十几厘米至几十厘米不等的草墩。由于冬季严寒，地面冻结，出现了网状的收缩，形成了凸起和凹槽。夏季，凸起部分生长出草并得到草根的包裹，更加坚实。周而复始，草根和泥碳长年累月凝结，就形成了塔头。图中这样大片的塔头地并不常见。

"Pagoda-head" is a metaphorical term that refers to grass mounds that elevate from the ground, ranging in height from a dozen to several dozen centimeters. As land is frozen in winter and becomes cracked in surface, it contracts in web-shaped patterns, leading certain parts to elevate whereas certain other parts to sink. When grasses sprout in summer, the elevated parts would be packed densely with roots and therefore become more consolidated. As time progresses, grass roots and peats would also accumulate, thus giving rise to what is known as the "pagoda-heads". Such an extensive stretch of pagoda-head land as shown in the photo is not something that can often be seen.

珍宝岛河滩草甸·黑龙江虎林

Riverbed meadows on the Zhenbao Island — Hulin, Heilongjiang

乌苏里江水在这片宽广平坦的河滩地上久久滞留，形成了大片的沼泽湿地、沼泽草甸和岛状森林草地，随着季节变换、水位消长的变化，它们也会相互转换，让你眼花缭乱，分不清这到底算湿地还是草地。

Waterflow of the Ussuri River linger as it passes this wide stripe of flat riverbed, leaving behind it a long string of swamps, marshy meadows and island-shaped forest grasslands. As the seasons evolve and water level fluctuates, these landscapes may even interexchange in bewildering manners, leaving one can't help but wondering whether they belong to wetlands or grasslands.

4-4 可持续发展

SUSTAINABLE DEVELOPMENT

暖性草地生态系统（含暖性草丛生态系统和暖性灌草丛生态系统）主要分布在本区山地降雨较集中的地方。草地植被对于防止水土流失具有重要作用，努力保护好暖性草地生态系统对于保障山区及山前平原地区的生态安全具有重要的意义。暖性草地生态系统经济功能的重要方面在于它是很好的放牧场与割草场。由于水、热条件较好，土壤肥力较高，本区牧草的蛋白质含量和适口性均较好，是营养丰富的牧草资源，单位面积载畜量较高。但长期以来，除了偏远的地方和深山区外，大部分地区此类草地生态系统都存在过度放牧、过度采伐、过度开垦利用等现象，草地普遍存在着不同程度的退化，所以，防治和修复草地生态系统的退化，保护好本区草地资源是非常重要的。

本区基本涵盖了属于中国自然动物地理区划中古北界东部的东北区、华北区，动物种类（尤其是两栖爬行类）相对贫乏，但东北区林草间生存着一些草食性或杂食性动物，如棕熊、梅花鹿、狍子、紫貂等，不少哺乳动物有冬眠现象或有储藏食物越冬的习性。松嫩平原的大量湿草地是丹顶鹤、白鹤等多种鹤类在迁徙过程中停歇和补充营养的地方。而华北地区野生动物物种较为稀少，尤其是平原地区，已经很难寻觅到较大体形的野生动物的踪迹了，相比起来，山区的野生动物状况会好一些。近些年来，山区常常发现有顶级物种华北豹的活动踪迹，说明有一定数量处于食物链下端的兔、鹿、麂等食草动物的存在，显示这里的自然生态环境有了较大好转。江苏大丰具有目前世界最大数量的大型食草动物麋鹿。

下一步要努力建设好以东北虎豹、松嫩鹤乡、辽河口、长白山、小兴安岭等涉草类和草原类的国家公园为主体的自然保护地体系，完善草原类的自然保护地空间布局和建设，加强中俄跨界自然保护地和野生动物生态廊道建设，构建好以自然保护地为重要支撑的祖国北方生态安全屏障体系。

Warm grassland ecosystems(including warm tussock ecosystem and warm shrubbery and tussock ecosystem) are mainly found in the mountainous areas of the region where rainfall is relatively concentrated. Given that grassland vegetation plays an important role in preventing soil erosion, it will be of great significance to keep the warm grassland ecosystems well protected, hence ensuring the ecological security of the mountainous areas and the plains that lie in front of the mountains. The economic value of warm grassland ecosystems lies in that they make up very good pastures and hay harvesting sites. Favorable water and heat conditions and higher soil fertility translate into relatively higher protein content and improved palatability of grasses, which in turn means nutrient-rich pasture resources and higher livestock carrying capacity per unit. However, except for those located in very remote places or deep in mountainous areas, most of the warm grassland ecosystems in this region have long been subjected to abusive uses — like overgrazing, excessive logging, exploitative land reclamation and

陡壁牧羊•山西阳泉

Sheep grazing at steep mountain slopes — Yangquan, Shanxi

太行山地降水集中，草地植被对于防止水土流失具有极为重要的作用，而这些地方的过度放牧和陡坡利用过度等现象使草地严重退化，尤其是在大于25度的陡坡上，水土流失不断加剧。因此，这些地方应大力实施退耕退牧还林还草，努力保护好这里珍贵的暖性草地和暖性灌草地，这对保障山区及山前平原地区的生态安全具有重要的意义。

In the Taihang Mountains where rainfalls tend to be highly concentrated, the role of vegetation in preventing water/soil erosion is especially important. Unfortunately, overgrazing and abusive use of steep mountain slopes in these areas have led to serious grassland degradation, particularly on slopes that are steeper than 25 degrees where water/soil losses are getting ever more serious. Therefore, efforts must be strengthened in these places to re-convert grazing lands and arable lands back to forests and grasslands, so as to better protect the precious warm grasslands and shrubbery-dotted grasslands that exist here, which will be of significant meaning to the eco-security of the mountainous areas and plains in front of them.

etc., leading in consequence to various degree of degradations. For this reason, it is imperative to take effective measures to prevent the grassland resources in this region from further degradation or to restore those that have already degraded.

This region basically covers all places that fall, in terms of China's natural zoogeographical zoning, within the north-eastern and northern China districts that lies in the eastern part of the palearctic realm. In spite of the relative insufficient presence of animal diversity (especially the amphibians and reptiles), there are plenty of herbivorous or omnivorous animals — for instance, brown bears, sika deer, roe deer, sable and *etc.* — inhabiting among the forests and grasslands in the northeast, where many mammals tend to hibernate or store food for winter. The vast stretches of wet grasslands dotted over the Songnen Plain are important places for migrating red-crowned cranes and whooping cranes to stop for rest or for nutrition top-ups. The diversity of wildlife species in the North China — especially on the plains — is comparatively poorer, with larger-sized wildlife species hardly seeable. The situation is relatively better in the mountains, where traces of flagship species like North China leopards have been found in recent years, indicating there are abundant supply of herbivores — like rabbits, deer, muntjacs and other — that sit at the lower end of the food chain, which in turn signifies notable improvements in local natural ecological environment. Dafeng in Jiangsu Province currently boasts the largest population of elks — a large herbivores — in the world.

The next step is to further optimize the spatial layout of rangeland-featuring natural protected areas through putting in place a sound natural protected area system that centers around the rangeland-featuring national park, covering the tiger/leopard habitats in the Northeast, the crane lands over the Songnen Plain, the mouth of the Liaohe River, the Changbai Mountain and the Xiaoxinganling Mountain Ranges. Efforts must also be strengthened to build an eco-corridor along the border of China and Russia, hence contributing to the formation of a reliable eco-security barrier at the north of the country.

牧羊归来•河南南阳

Back from sheep grazing — Nanyang, Henan

在华北平原农区中，草地只占很小的比例，老百姓饲养为数不多的牛羊作为"副业"，可以补充肉奶，增加收入，优化农业结构，但也必须要注意解决过度放牧和农牧冲突的问题，以保证农牧协调、可持续发展。

In the agriculture-dominant North China Plain where grasslands only take up a very small percentage of the land areas, local folks keep limited numbers of cattle and sheep only as a "supplementary industry" that — with its extra diary and meat outputs — brings some additional incomes and optimizes local agricultural structure. Nevertheless, due attentions must also be given to avoid over-grazing and to address the agriculture-husbandry conflict properly, making sure the two industries develop in well-coordinated and sustainable manners.

CHAPTER 5 第五章

南方山地丘陵草原区

The Southern Hilly Rangeland Region

南山牧场清晨•湖南邵阳
Morning on Nanshan Pastureland — Chengbu, Hunan

这里是中国南方最大的山地牧场，属典型的热性灌丛草地生态系统，跨湖南、广西两省（自治区）。这类次生生态系统是在亚热带森林遭到破坏后形成的，以次生、多年生草本为主，并保留一定数量原有植被中的乔木和灌木。植被受牲畜和人长期的不断干预，干扰度在一定范围内稳定波动，光、热、水条件又比较充分，使得这个次生生态系统相对稳定。这里也被列入全国首批12处红色草原之中。

Shown in the photo is the largest mountainous pasture land in areas bordering between Hunan Province and Guangxi Zhuang Autonomous Region in southern China, where the ecosystem belongs to that of typical hot shrubbery-grasslands. This is a secondary ecosystem that comes into being in the wake of destructed sub-tropical forests, where — though the predominant plants are secondary perennial herbaceous species — a certain number of tall trees and shrubberies from previous primary stands still survive. Though the vegetation has long been subjected to livestock-inflicted damages and human interference that fluctuate within a certain degree, favorable combination of heat, light and water still ensures that this secondary ecosystem is kept at a relatively stable state. It is also listed among China's first batch of red rangelands in honor of its glorious revolutionary legacies.

II
I
北京★
Beijing
IV
III
V
南海诸岛
China South Sea Islands
1:64 000 000
图例 Legends
I 内蒙古高原草原区
The Inner Mongolian Plateau Rangeland Region
II 西北山地盆地草原区
The Mountainous and Basin Northwestern Rangeland Region
III 青藏高原草原区
The Qinghai-Tibetan Plateau Rangeland Region
IV 东北华北平原山地丘陵草原区
The Northern and Northeastern Plain and Hilly Rangeland Region
V 南方山地丘陵草原区
The Southern Hilly Rangeland Region
国界 National Border
1：32 000 000

南方山地丘陵草原区指秦岭—淮河以南、青藏高原以东、横断山南部的华中、华南和西南的大部分地区，包括四川盆地、云贵高原、长江中下游平原、东南丘陵及台湾岛、海南岛等。这里属于中国的亚热带与热带气候区，其气候特点是气温较高，年平均气温大多在14~22摄氏度，雨量充足，年平均降水量大部分地区都在800毫米以上，最高可达2000毫米以上。其地形主要为低山丘陵，部分地区海拔可到2800米以上，为中山。土壤主要为黄壤、黄棕壤、山地红壤和砖红壤等，一般土层浅薄，有机质含量较低，土质黏重，结构不好，保水性能较差，肥力较低，土壤呈酸性反应。本区主要分布着各类热性草地生态系统和若干草甸生态系统类型。

本区以亚热带、热带的多年生草本植物为主，多系旱生植物与中生植物，间有少量灌木或乔木，其中很大部分是天然森林植被破坏后形成的相对稳定的次生草地生态系统。热性草丛生态系统与热性灌草丛生态系统，被统称为热性草地生态系统。在某些特定的地形与土壤分布区，由于干热的气候和人类活动的影响而形成了较稳定的干热稀树草原生态系统。

此外，本区还分布典型草甸生态系统和沼泽草甸生态系统。随着土壤和地表水分的进一步增加，就变成了沼泽湿地生态系统。沼泽湿地生态系统已属于湿地范畴（参见“一滴水生态摄影丛书”之一的《多样性的中国湿地》），本书不再涉及。

The Southern Hilly Rangeland Region covers a vast majority parts of Central China, South China and Southwest China that lie to the south of the Qinling–Huaihe line, to the east of the Qinghai-Tibetan Plateau, and to the south of the Hengduan Mountains — composed mainly of the Sichuan Basin the Yungui Plateau, the lower and middle reaches of the Yangtze River, the southeastern hills, as well as the Taiwan and the Hainan islands. Falling within the sub-tropical and tropical climate zones of the country, this region is characterized by comparatively higher temperature and abundant precipitations — with annual average temperature ranging between 14–22 Celsius degrees and annual average precipitation standing at 800mm, reaching as high as 2,000mm in some cases. The dominant landscape is low-lying hills, but medium mountains that stand over 2,800m above sea level are also present in certain areas. Soils in this region mainly include yellow soil, yellow-brown soil, mountainous red soil and brick-red soil that typically come with thin layers, low organic matter content, poor structure, low water retention rate, low fertility and slightly acid chemical properties. Widely distributed across this region are the various types of hot grassland ecosystems and a limited variety of meadow ecosystems.

This region is predominated with sub-tropical and tropical perennial herbaceous plants — mostly xerophytes and mesophytes — that are occasionally dotted with sparse shrubberies and tall trees. A vast majority of the ecosystems are relatively stable ones composed of secondary grasses, hot tussocks, and hot shrubbery-tussocks (which are known collectively as warm grassland ecosystems) that come into being in the wake of destructed natural forests. In certain areas that are endowed with some specific types of landforms and soils, fairly stable savanna-like sparse-tree rangeland ecosystem can also be found due to the influences of both dry and hot climates and human activities.

Ecosystems of typical meadows and marshy meadows can also be seen in this region. As the water content increases both in the soils and at the surfaces, the landscape gradually shifts towards marshy wetland ecosystems, which belongs to the category of wetlands (readers interested can refer to *Diverse Wetlands of China* that has previously been published under the " A Single Drop of Water Ecological Photography Album series").

5-1 热性草丛生态系统和热性灌草丛生态系统

HOT TUSSOCK ECOSYSTEM AND HOT SHRUBBERY AND TUSSOCK ECOSYSTEM

在中国热带、亚热带气候条件下，原有天然林在烧荒开垦、过度放牧以及其他长期不合理的人类活动和气候变化影响下逐步形成的多年生草本植物群落及动物、微生物群落一起与环境构成的有机整体，被称为热性草丛生态系统。若还伴有一定数量灌木或乔木的植物群落就被称为热性灌草丛生态系统。在中国，这两类草地生态系统主要分布在云南、贵州、四川、湖南、安徽、江西、福建、广东、广西、海南等省（自治区），并被统称为热性草地生态系统。其所处的地形主体为山地丘陵，土壤呈强酸性反应，有机质含量相对不高。

这里人为活动比较频繁，亚热带落叶、常绿阔叶混交林，常绿阔叶林和热带森林被人为破坏后，天然的森林生态系统被演替成次生生态系统，而热性草丛生态系统和热性灌草丛生态系统就是这种次生生态系统中的一种类型。由于光、热、水都十分充足，植物生长量和破坏力处于某种平衡状态，这类次生生态系统处于一个相对稳定的状态。如果人类活动停止，这一类生态系统有可能通过乔木和灌木的入侵，逐渐向与当地生态环境相适应的森林生态系统演变。

Natural forests that previously existed in China's tropical and sub-tropical areas, due to the long-term impacts of exploitative use by human beings — like reclamation through burning, over-grazing and *etc.* — as well as that of climate change, have gradually given way to integrated communities composed of perennial herbaceous plants, animals, microbial species and corresponding environment, which are referred to as hot tussock ecosystem. In cases where such communities are also joined by a certain number of shrubberies and tall trees, they will become hot shrubbery and tussock ecosystem. These two types of grassland ecosystems in China can mainly be found in provinces (autonomous regions) like Yunnan, Guizhou, Sichuan, Hunan, Anhui, Jiangxi, Fujian, Guangdong, Guangxi and Hainan, where the prevalent landform is hilly mountains that typically come with slightly acid chemical properties and low organic matter content.

As a result of frequent anthropogenic activities that reduce the subtropical deciduous and evergreen broad-leaved mixed forests, the evergreen broad-leaved forests and the tropical forests that previously flourished here into degraded states, the natural forest ecosystems are succeeded by secondary ecosystems, to which hot tussock ecosystem and hot shrubbery and tussock ecosystem belong. Given that the favorable combination of water, heat and light conditions is conducive to keeping the growth of the biomass and the damages at balance, these secondary ecosystems are generally very stable. Once the anthropogenic activities are checked, it is very hopeful that shrubberies and tall trees may intrude into this type of ecosystem, hence enabling it to evolve towards forest ecosystem that is better suited to local environment and conditions.

高山灌草地•福建武夷山
Alpine shrubbery grasslands — Wuyi Mountain, Fujian

位于南亚热带的武夷山上，大片的草坡草丛中生长着数量不少的灌木。秋天降临，五颜六色的灌木点缀在发黄的草丛中，坡下是翠绿的森林，相互映衬，别有一番南方山坡秋色风光。

Amidst the dense grasses growing on the slopes of the Wuyi Mountain located in southern sub-tropical belt, there exist huge number of shrubberies. As autumn sets in, the withered yellow grasses are often dotted with shrubs that come in a rich variety of colors, which forms a sharp contrast against the verdant forests further down the slopes and presents a picturesque autumn view.

石漠灌草丛·云南路南

Shrubbery grasslands over the stony desert — Lunan, Yunnan

在中国南方石漠地区，喀斯特地貌以无数、大大小小的石灰岩石镶嵌在灌草丛中最为典型。如此贫瘠的土地，仍然是老百姓常常放牧的地方。

In the stony desert areas of southern China that are well known for their karst landscapes, the most representative view is the shrubberies that grow among the randomly scattered limestone rocks that come in various shapes and sizes. Barren and bleak as the landscape is, it is still the place where local folks graze their livestock.

正在吃粽叶芦的野象·云南西双版纳

Wild elephants feeding on *Thysanolaena latifolia* — Xishuangbanna, Yunnan

大象是目前世界上最大的陆地动物，也是最大的食草动物，喜欢群居，主要栖息于热带雨林、季雨林及林间的沟谷、山坡、稀树草原及宽阔地带，常在海拔1000米以下的沟谷、河边、竹林、阔叶混交林中游荡。亚洲象喜食树叶、嫩芽、竹笋、野芭蕉和粽叶芦（又称大象草）等，更喜吃农作物，如香蕉和甘蔗等。

Elephants, the largest known terrestrial herbivore in the world that lives gregarious lives, mainly inhabit tropical rainforests, monsoon forests, and the valleys, mountain slopes, savanna rangelands and other open spaces situated amongst forests. They often roam over valleys, river banks, bamboo groves and broad-leaf mixed forests where the altitude is less than 1,000m. The Asian elephant (*Elephas maximus*) feeds mainly on leaves, tender shoots, bamboo, plantain and "elephant grasses" (*Thysanolaena latifolia*. But the more palatable to it are farmer-grown crops like bananas and sugar canes.

霍山石斛·安徽霍山

Huoshan dendrobium —
Huoshan, Anhui

石斛为多年生草本植物，被称为“中华仙草”，产于安徽省霍山县的霍山石斛是石斛中的极品，历史上一直贵为皇室专用。霍山石斛喜荫蔽，怕直射阳光，常与苔藓、石苇等植物附生在一起，其生长条件要求极为严苛，生长极慢，产量很低，因而特显珍贵。

The dendrobium, a perennial herbaceous plant, is often hailed as “the immortal grasses of China”. The Huoshan dendrobium(*Dendrobium huoshanense*) growing in Huoshan, Anhui Province is commonly credited as the best and used to be reserved for the exclusive uses of the royal courts. Being shade-prone and direct sunlight-aversive, this plant often appears side by side with mosses, *Pyrrosia lingua* and other plants that are highly demanding in growth conditions. Slow in incremental rate, its production is very small.

南山牧场牛·湖南城步

Cattle of the Nanshan pastureland — Chengbu, Hunan

南山牧场，全称为湖南省南山种畜牧草良种繁殖场，是中国南方最大的现代化山地牧场。南山是国家公园十大试点之一南山国家公园的组成部分，是中央红军长征翻越的"老山界"所在地，南山草原为我国公布的第一批"红色草原"之一。

The Nanshan Pastureland, whose full name is the Hunan Nanshan Breeding Site for Improved Livestock Fodder, is the largest modern mountainous pastureland in southern China. In addition to being a key composing part of the Nanshan National Park — one of the country's first 10 pilot national parks, Nanshan is also the former site of "Laoshanjie", a major milestone event during the Red Army's Long March, for which reason it is listed among the country's first batch of rangelands with revolutionary traditions.

休闲牛鹭图·台湾台东

Harmonious co-existence of cattle and egret — Taidong, Taiwan

草地上，一群黄牛在歇息，牛背鹭始终是它们形影相随的共生伙伴。牛背鹭喜欢跟随在耕田犁地的牛后面，啄食地里刚刚被翻出来的昆虫，或站在水牛或黄牛背上，啄食牛身上的寄生虫等。这类生物物种之间的共生现象，在自然界里常见。

A herd of cattle are lying leisurely on the grassland, accompanied by their forever-present partner — cattle egret (*Bubulcus coromandus*), a bird that likes to follow behind the cattle-drawn ploughs to feed on the insects excavated by the plough or to stand on cattle backs to peck on parasites living thereof. Such harmonious co-existence of different species is a very common view in nature.

黄草坝清晨·云南盈江

Morning at Huangcaoba — Yingjiang, Yunnan

在中国南方，叫草坝、黄草坝的地方不少。在热带、亚热带气候条件下，原生植被应该是森林，但由于人们砍伐森林、烧荒开垦、过度放牧，以及其他长期不合理的人类活动，逐步形成了草丛或灌草丛这样的次生生态系统。这种以多年生草本植物群落、灌丛及动物、微生物群落一起与环境构成的有机整体，被称为热性草地生态系统。

There are many places in southern China that are known as caoba or huangcaoba. The primitive vegetation of tropical and sub-tropical climate belts should have been forests. However, as a result of exploitative anthropogenic activities — like abusive logging, burning-based land reclamation, over-grazing and *etc.* — that had long been going on, forests that previously existed here have gradually given way to secondary ecosystems of tussocks or shrubberies and tussocks. The integrated body composed of perennial herbaceous plant communities, scrubs, animals, microbial communities and corresponding environment is generally referred to as the hot grassland ecosystem.

5-2 干热稀树草地生态系统

SAVANNAH-LIKE SPARSE-TREE GRASSLAND ECOSYSTEM

在热带与亚热带地区，有许多纵深的狭长峡谷，因“焚风效应”和“山谷风局地环流效应”而形成了干热的特殊生境。其特点是炎热而干旱的气候，贫瘠而浅薄的土壤，加上频繁的长期放牧、火烧、耕作等人类活动，形成相对稳定的干热稀树草地生态系统。这一生态系统是以多年生的耐旱草本植物为主，其间生长着耐旱灌木和稀疏孤立的乔木，从而呈现出特殊的群落结构与生态外貌。这类草地生态系统是热性草地生态系统的组成类型之一。

这类生态系统主要分布在横断山区的金沙江、澜沧江、怒江及南盘江、元江等的河谷山地等。其海拔一般都在1200米以下，这里旱季很长，气候炎热而干旱，年降水量虽可达1000毫米，但主要集中于雨季。在海南岛西部也分布有干热稀树草地。

In tropical and subtropical regions, there are many deep and narrow canyons where, due to the impacts of the so-called “burning wind effects” and “wind local circulation effects”, a very special type of dry and hot habitat comes into being. A defining characteristic of such places is that they tend to be extremely hot and dry in climate but barren and thin in soils. Plus the impacts of frequent anthropogenic activities such as grazing, burning and land cultivation, a relatively stable savannah-like sparse-tree grassland ecosystem often emerges. This ecosystem is predominantly composed of perennial drought-tolerant herbaceous plants that are sparsely dotted with some drought-tolerant shrubs and tall trees, thus presenting a special community structure and ecological view. This sort of ecosystem is a fundamental type among the hot grassland ecosystems.

This type of ecosystem mainly exists in the valleys of the Jinsha River, the Lancang River, the Nujiang River, the Nanpan River and the Yuanjiang River that flow among the Hengduan Mountains. Averaging less than 1,200 meters in altitudes, these places are very hot, and though the annual precipitation may reach 1,000 millimeters, they tend to be concentrated during rainy seasons, leading in consequence to fairly long and hot dry seasons. Besides, such savannah-like sparse-tree grasslands can also be found at the western parts of Hainan Island.

山水之间•云南德钦
Between mountains and rivers — Deqin, Yunnan

因“焚风效应”和“山谷风局地环流效应”，导致横断山三江并流区的诸峡谷中形成了干热的特殊生境，这里旱季漫长，炎热而干旱。图为金沙江峡谷的干热河谷稀树草地植被景观，近处为特征植物——仙人掌。

Due to the impacts of the so-called “burning wind effects” and “wind local circulation effects”, there emerge in the numerous valleys located among the Hengduan Mountains a very unique dry and hot habitat where the dry season is particularly long and dry. Shown in the photo is the savanna-like sparse-tree grassland landscape seeable in the Jinsha River Valley, the signature plant cactus is in he foreground.

▲ 干热河谷•云南元阳

Dry and hot river valley — Yuanyang, Yunnan

元江为红河在中国境内河段的名称，因河流多经过红色沙页岩地层，水呈红色，故得此名。元江河谷为云南红土高原上典型的干热河谷。

The Yuanjiang River — the section of the Red River before it flows out of China — passes through red sandy shale strata, which tints the water into a reddish color, hence named as such. The Yuanjiang Valley is a typical dry and hot one on the Red-earth Plateau of Yunnan.

▼ 稀树草地上的坡鹿•海南东方

Panolia siamensis roaming on the savannah-like sparse-tree grassland — Dongfang, Hainan

坡鹿在中国仅分布于海南西部的丘陵草坡地带，栖息于热带低海拔地势平缓的稀树草地、落叶季雨林中。图为冬季在大田稀树草地上生活的坡鹿母子仨。

The *Panolia siamensis*, a Level I protected fauna species of china found in China only on the hilly grassy slopes of western Hainan, typically live on low altitude gentle savannah sparse-tree grassland and deciduous monsoon rainforests in tropical areas. Shown in the photo are a doe and her two calves that are roaming leisurely on the Datian Savannah Sparse-tree Grassland in winter.

先花后叶的木棉•云南元江
Hysteranthous kapok —
Yuanjiang, Yunnan

木棉又称攀枝花，生长于海拔1400~1700米的干热河谷及稀树草地，也可生长在沟谷季雨林内。在干热地区，木棉先开花后长叶；在雨林气候条件下，花和叶共生。

Kapok, also known as Panzhihua in Chinese, grows mainly in dry and hot valleys and savannahs that lie 1,400–1,700 meters above sea level. It can also be found among monsoon rainforests located in river valleys. Whereas blossoms of kapok appear before leaves in hot and dry settings, the two tend to occur simultaneously under the rainforest climates.

5-3 草甸生态系统

MEADOW ECOSYSTEM

和各区都分布的草甸生态系统类似，本区的草甸生态系统以多年生草本植物为主，其环境特点是雨季有临时积水，土壤过分潮湿，经常通气不良，有机质不易分解，甚至可能有不同程度的泥炭积累，基本都分布在地势低洼、排水不畅的低地，如分布在河流两侧的河漫滩、低阶地、湖海周边、山间谷地或湿地外围等。其土壤类型为沼泽草甸土、泥炭草甸土等。

在本区，主要分布有典型草甸生态系统和沼泽草甸生态系统，在一些海拔较高的山丘地、水分丰富的地方，也分布有高山草甸生态系统。

Like that found in all other regions, the meadow ecosystem existing in this region is also primarily composed of perennial herbaceous plants. Defining features of the environment in which it exists can be summed up as follows: temporary waterlogging may occur during rainy seasons; the soils — made often of excessively humid swampy meadow soil and peat meadow soil — have poor aeration, making organic matters less likely to decompose and peat more prone to accumulate; it is mostly situated at low-lying and poorly drained places like riverside floodplains, low terraces, para-lake/sea settings, mountainous valleys, or the periphery of wetlands.

The meadow ecosystem prevalent in this region includes that of typical meadows and marshy meadows. The alpine meadow ecosystem can also occasionally be found in high altitude hilly places that has abundant water supplies.

高山草甸•湖南城步
Alpine meadows — Chengbu, Hunan

十万古田草甸是中国南方罕见的缓坡高山草甸，地势北高南低，海拔在1700米左右，因曾有古代人在此屯田，因此得名。这里是湖广古驿道的必经之路，更是沅江和珠江的发源地之一，对保护洞庭湖和珠江流域的水资源和生态系统安全有着十分重要的意义。

The Shiwangutian Meadow, rising about 1,700m in elevation and being higher in its northern part and lower in the southern, is a gentle slope alpine meadow that is not often seen in southern China. Far back in history, massive reserved troops were once stationed here and did farming work during peace times, thus getting its name, which literally meaning "ancient croplands that yielded 100 thousand bushels grains". In addition to being a strategic passage along the ancient Huguang Postal Route, it is also the origin of the Yuanjiang and the Pearl rivers. It plays a significant role in protecting the water resources of the Dongting Lake and the Pearl River catchment and in safeguarding their ecological security.

呦呦鹿鸣•湖北神农架

Mooing deer —
Shennongjia, Hubei

"呦呦鹿鸣，食野之芩。"芩是指生长在湿地草甸的禾本科植物。在神农架国家级自然保护区的大九湖片区，梅花鹿群在湖周草甸中穿梭往来，此情此景让人仿佛也感受到了《诗经》的那种意境。

A widely quoted line from the *Book of Songs* depicts the scene when deer feeds leisurely on its favorite Gramineae plants. Herds of Sika deer roam carefree amidst the meadows in the Dajiuhu area situated within the Shennongjia National Nature Reserve, bringing about a vivid representation of the view depicted in the poem.

勿忘草•重庆涪陵

Myosotis alpestris —
Fuling, Chongqing

勿忘草为多年生草本植物，喜生长于山地林缘或山坡、山谷草地处，其适应力强，喜干燥、凉爽的气候，忌湿热，喜光耐旱。勿忘草的花语为永恒的爱、永远的回忆，也许意由草名而来。

Myosotis alpestris, a highly adaptive perennial herbaceous plant that thrives under cool and dry climate, often grows on forest edges or on grasslands over mountain slopes and valleys. It symbolizes eternal love and lasting memories, which presumably is just what its name implies.

广布小红门兰•贵州黔东南

Ponerorchis chusua —
Qiandongnan, Guizhou

广布小红门兰为兰科植物中的一属，多年生草本。花多呈紫红色或粉红色，花朵不像很多兰花那样偏向一侧而多向，生于海拔630~3800米的山坡、林下、灌丛下或草甸上，以西南高海拔山地中生长数量居多。

Ponerorchis chusua a perennial herbaceous plant that belongs to the family of Orchidaceae. Mostly pink or purple in color, the blossoms of this plant, unlike that of many other orchid species, tend to grow in multiple directions. It often grows on mountain slopes, in forests or shrubbery understories, or over meadows that range between 630–3,800m in altitude, particularly in the high altitude mountainous areas of the southwest.

塔头沼泽•
湖南邵阳
"Pagoda-head" marsh — Shaoyang, Hunan

十万古田系明、清两朝瑶民在这里耕作、聚落的遗址，这里具有低等苔藓、草本植物、木本植物等完整植被演替系列，是沅江和珠江的发源地之 ·。图为十万古田的高山塔头沼泽草甸。

Shiwangutian, the former settlement of the Yao people during the Ming and Qing dynasties, showcases the complete process of plant succession — ranging from mosses, herbaceous plants all the way through woody plants. It is also one of the sources from where the Yuanjiang River and the Pearl River originate. Shown in the photo is the alpine "pagoda-head" marshy meadow located at Shiwangutian.

高原鹤之家•
云南昭通
Home of plateau cranes — Zhaotong, Yunnan

云南大山包的大海子周边是大片的典型草甸和沼泽草甸，冬天来临，号称高原鹤的黑颈鹤就大批地迁飞到这里。每年黑颈鹤在这里生活的时间长达176~202天，这里是它们理想的越冬栖息地。

Lying around the Dahaizi, a lake tucked among the Dashanbao Mountain, are vast stretches of typical meadows and marshy meadows, where black-necked crane — hailed as the plateau cranes — would arrive in flocks each year as winter advents. The time span that black-necked cranes stay here adds up to 176–202 days each year, making this place a most ideal wintering habitat for this precious fowl species.

5-4 可持续发展

SUSTAINABLE DEVELOPMENT

热性草地生态系统因处于热带、亚热带地区，总体上具有很大优势，这些地区热量充沛、降水多、植物种类丰富，且生长季长、生产力高。但是，也有许多不利之处，其大多分布在不同海拔的山地，有的山势陡峻且土层浅薄，土质黏，有大小不等的石砾，土壤保水能力弱等，因此，一遭降雨则水土流失严重。而且，这些地区气温高，物质分解快，植物必需的营养元素缺乏，土壤一般呈强酸性反应，总体上看，土壤肥力不高。再者，其大部分地区长期不合理的人类活动，如过度放牧或开垦，导致自然植被遭到不同程度的破坏，生态系统退化严重。另外，这些草地生态系统大部分都是次生生态系统，如果能“师法自然”，充分利用这里优良的光、热、水条件，下力气采取保护与生态修复的科学措施，完全可以让退化草地向着良性演替的方向转化。

热性草地生态系统的植物种类组成复杂，具有群落及类型的多样性和特殊的价值，有很多种类都具有观赏价值和药用价值。在有条件的地区，可以充分利用其优势，发展特色草种植业（如兰草等观赏草、花卉等），很有市场前景。

本区基本涵盖了属于中国自然动物地理区划中东洋界的华中区、华南区和西南区的东南部分，这里是中国最南，最热，降水量最充沛，光、热、水条件都非常优越，生物多样性异常丰富的地区，分布着羚牛、海南坡鹿、多种灵长类、亚洲象、绿孔雀等数量少且分布狭窄的食草性动物或杂食性动物，以及野果草、黄茅、芒萁、粽叶芦等野生植物及多种兰科植物。

中国南方的草地面积占比很小，基本都呈小块、破碎、杂乱的状态，容易被人忽视。虽然也存在有个别较大片的草地，但往往处在比较偏远的丘顶、山间，数量也不多。由于这里光、热、水条件俱佳，加上人口密集，这些草地绝大多数都分布在地头、村后、坡上、山中，与老百姓的生产、生活有着密不可分的关系。这些分散的草地往往给耕地的牛、拉车托物的马、供肉食的牛和羊等牲畜提供了最方便的采食和休息地，尽管这里的牲畜大都离不开人工补饲，但这种不起眼的小块草地，却是老百姓日常生活离不开的宝地。在人均耕地面积如此紧张的情况下，这些草地能够保留下来是非常不容易的，肯定也是有其客观必然性的，我们应当充分认识到这些草地的重要性，保护好它们，勿以善小而不为。

南方发展草产业，除了传统意义的肉、奶外，发展兰草等观赏草、花卉等产业也有着特殊的意义，观赏鲜花、药用石斛等不仅有其经济价值、生态价值，还有其独特的文化价值。

下一步要努力建设好以武夷山、海南热带雨林、南麂列岛—北麂列岛、钱江源—百山祖、黄山（牯牛降）、井冈山、神农架、张家界、南山、南岭、丹霞山等涉草类和草原类国家公园为主体的自然保护地体系，完善草原类自然保护地空间布局和建设。做好与周边国家的跨境保护和野生动物生态廊道建设，构建好以自然保护地为重要支撑的生态安全屏障体系。

收工回家•江西赣南

Going home upon completing the day's work — Gannan, Jiangxi

图为农民收工后，将放在村后草坡上吃草的黄牛一起带回家。放在草坡上的牛，往往得用一根长绳固定在一根随处可钉可拔的木桩上，这样既能让牛多换地方并转着圈地吃草，又能避免牛丢失或乱跑吃了别人家的庄稼。

Shown in the photo is the sight when a local farmer, upon finishing his work in the field, is picking up the cattle that has been left to graze on the grassy slope behind the village and going home together. The cattles grazing over the slope is often fastened to a long rope, the other end of which is tied to a wooden stake that can be readily nailed into the ground and wrenched from the ground. This will not only allow sufficient space for the cattle to forage, but also prevent it from going astray or running accidentally into the croplands of other folks and ruin their crops.

Existing in tropical and sub-tropical regions that is endowed with abundant heat and rainfall, rich plant diversity, long growing seasons and higher productivities, the hot grassland ecosystem is, generally speaking, situated at a very advantageous position. Nevertheless, there are also some disadvantages, one of which is that it could be highly vulnerable to serious water/soil erosions once heavy rain strikes, given that this sort of ecosystem is mostly located in mountainous areas where the steep slopes tend to be covered with shallow layers of sticky and gravel-strewn soils that have low water-retention capacities. Moreover, due to the high temperature and fast decomposition, essential nutrients needed by plants are often in deficit in the slightly acid soils that typically exist here, leading in result to comparatively lower soil fertility on the whole. To make it even worse, most parts of these areas have long been exposed to irrational anthropogenic activities — like over-grazing and unchecked land reclamation — that have more or less damaged its natural vegetation and resulted in grave ecosystem degradation. In addition, given the presence of highly favorable water, heat and sunlight conditions here, so long as the "go-after-nature" philosophy is strictly observed in our efforts to protect and restore the degraded ecosystems, it is very likely that these grassland ecosystems — mostly secondary ones — can be put back on to the track of virtuous succession.

The hot grassland ecosystems, with its highly complex plant composition, is often highly valuable for the rich diversity and uniqueness of its plant communities — many species being in possession of extraordinary ornamental and medicinal values. In places where condition permits, attempts may be made to take the fullest advantages of such ecosystem through developing some special grass-based plantations (that of ornamental orchid and other flowers, taking for instance) that boast promising market potentials.

This region falls basically, in terms of China's natural zoogeographical zoning, within the Central China, South China and southeastern part of the Southwest China districts of the Oriental realm. Being the southmost and hottest parts of China that are endowed with most abundant rainfalls, most favorable combination of water, heat and light conditions, and most rich biodiversity, it is the home of some precious and rare herbivores and omnivores that are only narrowly distributed here — takins, *Rucervus eldii*, several variants of primates, Asian elephants, blue peacocks and *etc.* for example, as well as the home of many precious floral species of the Orchidaceae and other families— *Amomum koenigii, Heteropogon contortus*, *Dicranopteris pedate* and *Thysanolaena latifolia*, to name just a few.

Grasslands in southern China take up just a very small proportion of the land area — typically existing in the form of highly fragmented and randomly-scattered patches — and therefore tend to be often ignored. Though there occasionally also are some large grasslands, they are generally found only in some extremely remote and isolated mountains or hilltops, and are limited in numbers. Because of the presence of favorable water, heat and sunlight conditions, places in southern China where grasslands exist are often also population-dense areas, for instance around croplands, behind villages, on slopes or in mountains, all of which are closely tied with the daily life and productive activities of local communities. These fragmented grasslands provide easily accessible foraging and rest sites for both beasts-of-burden (like oxen and horses) and meat-providing animals (like cattle and sheep). Although all these livestock in this region rely to a large extent on humans for their sustenance, these humble-looking small plots of grasslands make up indispensable properties that local folks have for raising their livestock. Given the extremely limited availability of per-capita arable lands, it hasn't been easy for the local people to have kept these grasslands from being reclaimed or used otherwise. We must be fully aware of the objective needs for their existence as well as of the critical roles they play, and take measures to keep them well protected, rather than merely dismiss them off as being negligible only because of their seemingly humble state.

Besides their traditional values in providing meat and milk, developing grass-based industries (for instance orchid cultivation and other ornamental flower nurseries) in the south is also of significant meaning in other aspects. In addition to the economic and ecological values generated by ornamental flowers and medicinal *Dendrobium* spp., distinctive cultural values can also be obtained from such endeavors.

The next step of our protective efforts in this region should be placed on building the nactural protected area system that centers around the rangeland-featuring national park covering the Wuyi Mountain, the Hainan Tropical Rainforests, the Nanji and Beiji Islands, the Qianjiangyuan–Bashanzu, the Huangshan Mountain (Gu'niujiang), the Jinggang Mountain, Shennongjia, Zhangjiajie, Nanshan Mountain, Nanling, the Danxia Mountain and *etc.*, hence further optimizing the spatial layout of natural protected areas serving this purpose. Efforts must also be strengthened to put in place cross-border eco-corridors for the transnational protection of wildlife with neighboring countries and to lay down solid foundations for the nation's ecological security barrier.

物种名录

LIST OF SPECIES COVERED IN THE BOOK

为了便于读者查阅和进一步了解物种情况，此部分专门列出了在本书中出现的物种（按物种出现的页码排序）及其中文名、学名、保护级别、濒危程度等，并标记了该物种在书中出现的页码。

注1：为保护野生动植物，维护生态平衡，我国先后颁布了《中华人民共和国野生动物保护法》和《中华人民共和国野生植物保护条例》，规定国家对珍贵、濒危的野生动植物实行重点保护，并根据物种的珍贵濒危程度和管理严格要求依次分为国家一级保护野生动植物和国家二级保护野生动植物（以下简称国家一级、国家二级）。本书中的野生动物的国家保护级别依据为《国家重点保护野生动物名录（2021）》，野生植物的保护级别依据为《国家重点保护野生植物名录（2021）》，分别简写为CHINA Ⅰ、CHINA Ⅱ。

注2：世界自然保护联盟（简称IUCN）编制的《世界自然保护联盟濒危物种红色名录》（简称《IUCN红色名录》）是被全球各国广泛接受和使用的受威胁物种分级标准体系，将物种编入9个不同的保护级别，依次为：灭绝（EX）、野外灭绝（EW）、极危（CR）、濒危（EN）、易危（VU）、近危（NT）、无危（LC）、数据缺乏（DD）和未予评估（NE）。本书中物种的IUCN保护级别参考网站为《IUCN红色名录》官方网站。

注3：《濒危野生动植物种国际贸易公约》（简称CITES）是全球缔约国之间为了保护野生动植物物种不至于由于国际贸易而遭到过度开发利用而进行的国际合作。CITES将受管理的野生动植物物种按照其物种状况及其受贸易影响的严重程度依次列入附录Ⅰ、附录Ⅱ和附录Ⅲ。本书中物种的CITES保护级别依据2023年版的《濒危野生动植物种国际贸易公约》，分别简写为CITES Ⅰ、CITES Ⅱ 和CITES Ⅲ。

To provide readers with easily accessible reference materials and further details concerning the wildlife species covered in the book, a complete list (sorted in alphabetical order of their page numbers) is specially prepared, including such data as their Chinese names, Latin names, protective status, vulnerability degree, together with the page numbers in which they appear.

Note 1: In order to protect wildlife and keep its ecological balance, China consecutively promulgated the *Wild Fanna Species Protection Law of the People's Republic of China* and the *Regulations of the People's Republic of China with Regard to Wild Flora Species Protection*, which clearly stated that targeted protections shall be given to precious and threatened wildlife species by classifying them into 2 categories—*List of Level I Protected Wildlife Species of China and List of Level II Protected Wildlife Species of China* (abbreviated in the following respectively as the Level I and Level II lists) — and subjecting them to rigorous protective measures in accordance with how valuable they are and to what extent they are threatened. So far as this current book is concerned, the protective status given to each fauna and flora species is based *on the National List of Key Wild Fauna Species to Be Protected (2021)* and the *National List of Key Wild Flora Species to Be Protected (2021)* respectively, abbrevated as CHINA I, CHINA II.

Note 2: *The IUCN Red List of Threatened Species* (abbreviated as the Red List) compiled by the International Union for Conservation of Nature (IUCN), which is widely recognized and adopted by the international community as the standard system, classifies wildlife species into 9 different categories, namely: Extinct (EX), extinct in the wild (EW), critically endangered (CR), endangered (EN), vulnerable (VU), near threatened (NT), least concern (LC), data deficient (DD), and not evaluated (NE). The IUCN protective status used in this book is based on information retrievable from the official website of the *Red List.*

Note 3: *Convention on International Trade in Endangered species of Wild Fauna and Flora* (CITES) is an international agreement signed by all its COPs (Conference of Parties), which aims to ensure that international trade in wild animals and plants does not threaten their survival. Depending on their current status and the extent to which they are impacted, all wild animal and plant species covered by CITES are respectively included in Appendix I, Appendix II and Appendix III. The protective status used in the present book are all based on *CITES 2023,* abbrevated as CITES I, CITES II and CITES III.

本书中鸟类、兽类、爬行动物和植物名录分别参考以下文献。

蒋志刚.中国哺乳动物多样性：编目、分布与保护[M].福州：海峡书局,2024.
王剀,任金龙,陈宏满.中国两栖、爬行动物更新名录[J].生物多样性,2020,28(2):30.
郑光美.中国鸟类分类与分布名录[M].第四版.北京：科学出版社,2023.
中国科学院.植物科学数据中心[EB/OL].(2020-09-01)[2024-07-01].https://www.plantplus.cn/cn.

Lists of aves, fauna, reptiles and flora covered in the book are based on the following materials.

JIANG Zhigang. The Diversity of Mammals in China: Index, Distribution and Protection[M]. Fuzhou: Straits Publishing House Co., Ltd., 2024.
WANG Kai, REN Jinlong, CHEN Hongman. Updated index of amphibian and reptile animals in China [J]. Biodiversity, 2020, 28(2):30.
ZHENG Guangmei. Classification and Distribution of Birds in China (the 4th edition) [M]. Beijing: Science Press, 2023.
Chinese Academy of Science. Plant Science Data Center [EB/OL].(2020-09-01)[2024-07-01].https://www.plantplus.cn/cn.

猫头刺　P13
Oxytropis aciphylla

柳兰　P43
Chamerion angustifolium

麻花头　P43
Klasea centauroides

黑琴鸡　P44
Lyrurus tetrix
CHINA I　LC

金莲花　P49
Trollius chinensis

田鹨　P50
Anthus richardi
LC

雪鸮　P51
Bubo scandiacus
CHINA II　UV

针茅　P51
Stipa capillata

蒙原羚　P53
Procapra gutturosa
CHINA I　LC

草原雕　P53
Aquila nipalensis
CHINA I　EN

石鸡　P59
Alectoris chukar

细枝羊柴　P60
Corethrodendron scoparium

碱蓬　P66、P196
Suaeda glauca

薹草　P67
Carex breviculmi

白臀鹿　P80
Cervus elaphus macneilli
CHINA I　LC

芨芨草　P82
Neotrinia splendens

肉苁蓉　P84
Cistanche deserticola
CHINA II

阿勒泰盘羊　P85
Ovis ammon
CHINA II　NT　CITES II

蒙古野驴　P86
Equus hemionus
CHINA I　NT　CITES I

鹅喉羚　P86
Gazella subgutturosa
CHINA II　EN　CITES I

大鵟　P87
Buteo hemilasius
CHINA II　LC

金雕　P92
Aquila chrysaetos
CHINA I　LC

多枝柽柳　P93
Tamarix ramosissima

骆驼刺　P94
Alhagi camelorum

白刺　P95
Nitraria tangutorum

野骆驼　P96
Camelus ferus
CHINA I　CR

野马　P96
Equus ferus

盐角草　P105
Salicornia europaea
CHINA I　EN　CITES I

马尿泡　P116
Przewalskia tangutica

总状绿绒蒿　P118
Meconopsis racemosa

唐古特雪莲　P119
Saussurea tangutica

野牦牛　P120
Bos mutus
CHINA I　UV　CITES I

独一味　P124
Phlomoides rotata

棕熊　P126
Ursus arctos
CHINA II　LC　CITES I

绿尾虹雉　P126
Lophophorus lhuysii
CHINA I　VU　CITES I

岩羊　P131、P172
Pseudois nayaur
CHINA II　LC　CITES III

藏野驴　P132、P154、P157
Equus kiang
CHINA I　LC　CITES II

西藏杓兰　P136
Cypripedium tibeticum
CITES II

蓝玉簪龙胆　P137
Gentiana veitchiorum

黑颈鹤　P138
Grus nigricollis
CHINA I　NT　CITES I

高原鼠兔　P139
Ochotona curzoniae
LC

红花绿绒蒿　P140、P141
Meconopsis punicea
CHINA II

隐蕊杜鹃　P141
Rhododendron intricatum Franch.

柯氏鼠兔　P148
Ochotona koslowi
EN

高山兀鹫　P149
Gyps himalayensis
CHINA II　NT

普氏原羚　P150、P175
Procapra przewalskii
CHINA I　EN

藏羚　P151、P157
Pantholops hodgsonii
CHINA I　NT　CITES I

藏原羚　P155
Procapra picticaudata
CHINA II　NT

高原兔　P159
Lepus oiostolus
UV

羽毛三芒草　P159
Aristida pennata

无斑山溪鲵　P163
Batrachuperus karlschmidti
CHINA II　UV

祁连圆柏　P163、P166
Juniperus przewalskii

红隼　P164
Falco tinnunculus
CHINA II　LC　CITES II

西藏马鹿　P166
Cervus elaphus
CHINA I

马麝　P167
Moschus chrysogaster
CHINA I　EN

小叶金露梅　P170
Dasiphora parvifolia

银叶杜鹃　P170
Rhododendron argyrophyllum Franch.

峨眉冷杉　P170
Abies fabri(Mast,)Craib

驯鹿　P183
Rangifer tarandus
UV

牛皮杜鹃　P187
Rhododendron aureum

梅花鹿　P187
Cervus nippon
CHINA I　LC

东北马鹿　P188
Cervus canadensis xanthopygus
CHINA II

大鸨　P190
Otis tarda
CHINA I　UV　CITES II

狍　P193
Capreolus pygargus
LC

麋鹿　P200
Elaphurus davidianus
CHINA I　EW

亚洲象　P210
Elephas maximus
CHINA I　EN　CITES I

霍山石斛　P211
Dendrobium huoshanense
CHINA I　CR　CITES II

牛背鹭　P213
Bubulcus coromandus
LC

坡鹿　P216
Panolia siamensis
CHINA I　EN　CITES II

勿忘草　P220
Myosotis alpestris

广布小红门兰　P221
Ponerorchis chusua
CITES II

后 记

POSTSCRIPT

草原对我来说，是一个既陌生又不算太陌生的领域。说陌生是因为原林业部（国家林业局）职能所涵盖的“三个系统一个多样性”只包括了森林、湿地、荒漠三个生态系统和野生动植物这一个生物多样性，并不包括草原。说不算太陌生，是因为我几十年来一直从事“三个系统一个多样性”的调查监测、科研教育和行政管理工作，以及在做全国自然保护区的建设与管理工作时，往往都要和草原、草地打交道。

我们无论是从科学上讲还是在实践中行，森林、湿地、荒漠生态系统和草原生态系统都是割不断、理还乱的关系，如森林里有草地、草地中有森林，森林分类中有灌木林类型、草原分类中有灌草地类型；又如，有水有草的地方，可以被划为湿地，也可以被划为草地；再如，荒漠中有草原荒漠，草原里有荒漠草原，等等。要划建和管理一个自然保护区，往往在这个自然保护区里，森林、湿地、荒漠、草原等各类生态系统都有。自然保护区作为一个整体生态系统保护的最有效手段，你能说这个被称作森林类型的自然保护区只保护森林，而其中的湿地、荒漠、草地都不保护？野生动物更是如此，它们在森林内栖息，在草地中觅食，在湿地里饮水，在荒漠上撒欢。我在几十年的自然保护工作中，少不了与草原打交道，因此，说对于草原不算太陌生还是有底气的。

2018年，国务院机构改革，原来由农业部畜牧司草原处管理草原的职能划归国家林业局，国家林业局组建为国家林业和草原局。我理解，这绝对不是政府部门管理分工上简单的拆拆并并，而是草原管理职能的重大改变，是从重草原的生产功能到生态与生产功能并重的根本性改变，是赋予草原国家重要生态屏障地位的重大提升。

为了深入了解草原，在原有的基础上，我除了恶补草原方面的科学知识外，还自费天南海北地去草原进行考察，向各方面的草原专家学习请教。为了更实际客观地了解中国草原，我还专门到俄罗斯、蒙古国、哈萨克斯坦、乌兹别克斯坦、厄瓜多尔以及澳大利亚、新西兰等草牧业发达的国家进行考察学习和收集资料。

在考察学习的过程中，我深刻体会到，草原或许是陆地生态系统（森林、湿地、荒漠、草原、农田）中最不起眼、最难辨识的生态系统，草原没有森林那样雄壮伟岸、外观特征明显；没有湿地那样迂回流转、界限分明；也没有荒漠那样粗犷奇形、印象深刻；更没有农田那样精耕细作、大地绣花。相比之下，草本植物微小的生命往往因被踩在脚下而不受人重视。但是，草太伟大了，渺小的草能够聚少成多、集腋成裘，无数的小草聚在一起，居然就能铺天盖地，宽广无垠得像波涛大海一样，无微不至得像柔软皮肤一样，时刻保护着我们的地球。草原是中国陆地覆盖面积最大的绿色植被，无论是在森林之上、湿地当中，还是在荒漠深处，都有草的行踪。贫瘠干旱的石头缝里有草的身影来证明生命的伟大，肥沃湿润的土地上有“野火烧不尽，春风吹又生”来证明草生命的坚韧不拔。草是那么渺小，其生命力又是如此强大，在地球生命系统的大循环中，草是初级的生产者，更是位卑未敢忘忧国的英雄，它从来不出风头闹显摆，始终默默地做好自己！我以为，人类用什么样的词句来赞美草都绝对不会过誉。人类应该充分尊重草原，并在更高的层次上重新认识草原。

就保护森林生态系统而言，为了保护森林不受或少受人为干扰，我们建立了很多禁伐区、天然林保护区及多种类的自然保护地，希望森林在少受或不受人类干扰的情况下，保持或恢复发展到森林的顶级群落，使各种动植物、微生物在其中自由生存，维持着良好的生态平衡，成为一种生态位最完善、内部结构最完整、生物链最全面的系统。可以这么说，生态功能最完善的森林生态系统是人为干预越少越好的生态系统（这里不含为生产木材而建设的人工商品林）。我们不仅要保护现有的天然林，还要努力培育一些次生林，甚至人工林，进而发展到生态林，努力使之向顶级群落转化，就是因为森林是陆地生态系统的主体，人类及野生动植物最需要的是它天然的、不可替代的、综合的生态功能，这是一个必须努力保持原真性的生态系统。

就农田生态系统而言，人们要精耕细作，从土地的耕翻施肥、选种育苗、抚育成长直到收获整个过程的每一个阶段，人的关怀和干预都无微不至。农田生态系统就是一个自始至终充满了人为干预的生态系统，人为干扰越精细、越科学，农田生态系统就越好。人类培育农田的目的是为了多产粮食，农田生态系统是人类生活必须依靠的，是一个纯生产性的生态系统。

既不同于人工干预越少越好的森林生态系统，也不同于人工干预越多越好的农田生态系统，草原生态系统是一个草（含野生植物）—畜（含野生动物）—人共生的生命共同体（以下简称草—畜—人共生），是兼有生产、生态、生计三个功能的生态系统，缺一不可。当然，我们这里说的草原生态系统是指以畜牧业为基础的生态系统（没有畜牧业的原始草原现在极少，专门用于打草的人工草场也不在本书讨论之列）。

我们讨论的这个草原生态系统是一个“草—畜—人”相互影响、相互制约、动态平衡的生态系统，这个草原生态系统人工干预多了不行，人工干预少了也不行，而现在我们草原的大面积退化、沙化就是人工干预过度，动态平衡被人为强行打破的结果。

草原的“生产者”是绿色草本，包括小灌木等野生植物（以下统称“草”），它们被“初级消费者”——牛、羊、马、骆驼等家畜及羚羊、野牦牛等野生动物（以下统称“畜”）采食大约2/3，除了通过消化吸收其中的53%之外，还有35%转化为粪、尿进入草原土壤。然后，兽、猛禽、蛇等“次级消费者”通过捕食“初级消费者”进行能量转换。最后，经过“分解者”——微生物及某些土壤生物转化，其中97%的物质与能量最终被草吸收利用。在这个生命循环系统中，“草”是草原的第一生产力，“畜”是草原的驱动力，“人”既是“次级消费者”，更是“促进力”或“破坏力”。三者中“人”是影响力最大的一方，当“草”生长与“畜”采食在“人”的调控下达到动态平衡时，草原生态系统就会不断更新，实现可持续利用，从而进入良性循环。反之，草原生态系统就会遭到破坏，甚至崩溃。

认识草原，我们不能只讲畜牧不讲生态，也不能离开人、畜去讲所谓“理想”的草原；不能以牛羊多少论成败，也不能以无牛羊为理想。我们必须清楚地认识到，草原生态系统是一个“草—畜—人”相互影响、相互制约、动态平衡的生态系统，一个与森林、湿地、荒漠、农田、海洋都不相同的且极为特殊的生态系统。

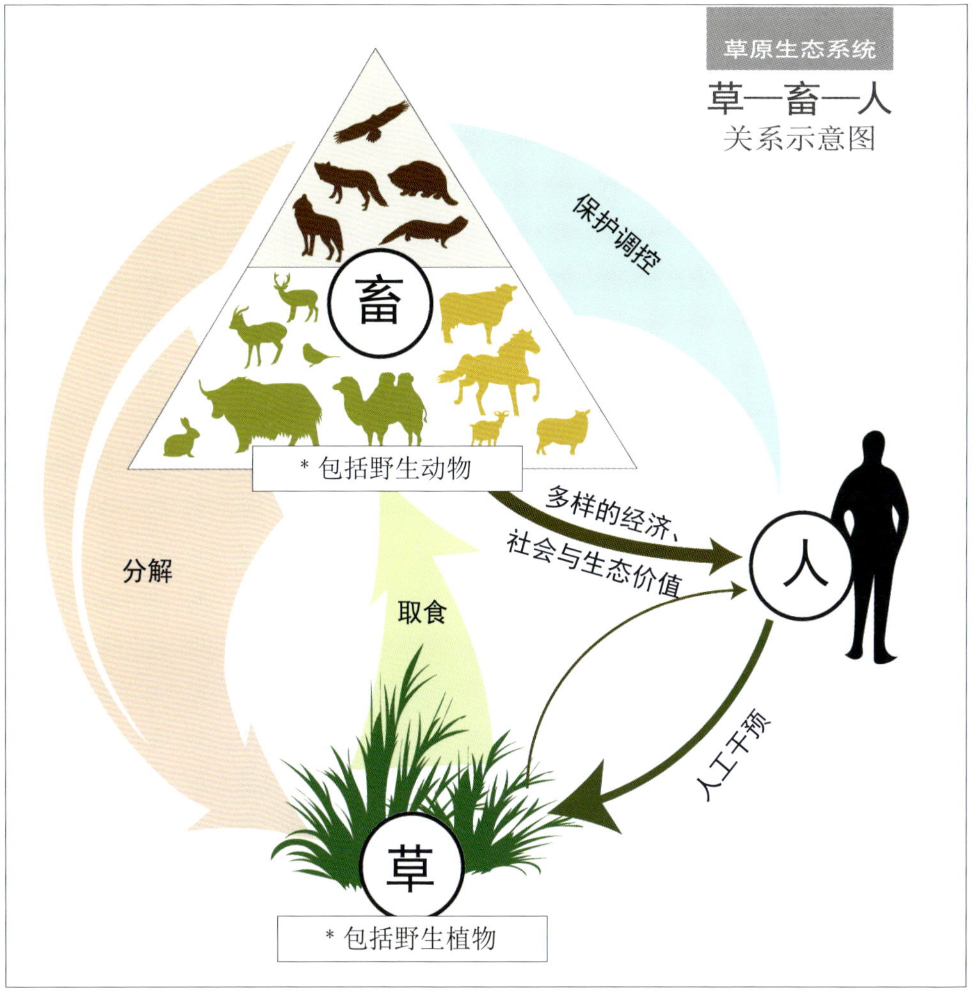

草虽小却种类繁多，加上它分布类型多样性，还有它的水平地带性、垂直地带性，加上高原地带性的分布，更让人眼花缭乱。用生态摄影这个崭新手段来讲述中国草原的故事，是呈现多样性的中国草原非常大的难题。既要用尽量通俗的语言来讲清楚那些绕来缠去的不同草原类型，同时还要找到辨识特征并不像其他生态系统那么明显的草原类型图片，这是我在过去编写中国森林、湿地、荒漠生态系统书时未曾遇到过的难题。我唯有以勤补拙。近三年起起伏伏的新型冠状病毒肺炎疫情又让一个个实地考察计划中止，我只能让自己整天钻在书中和埋进资料堆里，翻阅和学习大量的科技资料，在过去积累的百万张以上的照片库里去寻觅，如同大海捞针。当然，这样做的好处是，边学习边整理边编写的过程让疫情之后的实地补拍更有针对性，提高了工作效率。

我作为自改革开放以来中国现代自然保护事业这段历史的见证者和亲历者之一，无论是在工作期间，还是退休至今，始终是这个事业最热心的参与者和最忠实的追随者，对于这种人生命运的安排，我感到无比自豪和骄傲。

在几十年的自然保护工作中，我有一个非常深刻的感悟，就是中国的自然保护事业仅靠政府的努力推动和科学家的不断呼吁是远远不够的，必须依靠广大人民群众自然保护意识的觉醒、提高和实际的行动才能实现！因此，我暗暗下定决心，希望通过自己的不懈努力，向社会、向大众推出一套反映中国生态系统的科普丛书，来宣传群众、鼓舞群众，为提高大众的自然保护意识，作出自己应有的微薄贡献！

自2007年起，我就开始了“一滴水生态摄影丛书”的酝酿和创作，时至今日也一直没有停下来。刚开始时，也不知道仅凭个人的知识和努力能不能完成整个系列，我写下了这样一句话来不断鼓励自己：“如果一滴水能够反映太阳的光辉，那么无数滴呢？”出了一本，系列就多一本，就离终点近了一些。

“小车不倒只管推，推到哪里算哪里。”我就这么暗暗下定决心“撸起袖子加紧干”。2008年，“第一滴水”——《一滴水生态摄影集》出版；2011年，“第二滴水”——《多样性的中国森林》出版；2014年，“第三滴水”——《多样性的中国湿地》出版；2017年，“第四滴水”——《多样性的中国荒漠》出版；2021年，“第五滴水”《我们在中国——多样性的中国野生动物》出版；2022年，“第六滴水”——《全景湿地》出版；“第七滴水”——《多样性的中国草原》即将出版。至此，我不仅已经完成了要出好、出全“三个系统一个多样性”的夙愿，还在国家林业局改名为国家林业和草原局之后，赶上了时代需要的步伐，于2025年推出了丛书的最后一部——《多样性的中国草原》。

至此，历时17年完成了7部书，中国陆地生态系统的全部自然生态系统（森林、湿地、荒漠和草原）和中国野生动物均已圆满完成，我感到甚为欣慰。这些书中，有好几部书分别成为了中国政府送给北京奥运会、联合国国际公约缔约方大会或国际组织的礼品，有的还获得了中国科普作家协会优秀科普作品金奖、科技部全国优秀科普作品、梁希科普作品类奖等奖项，同时也获得了若干国内、国际各种媒体和友人的广泛赞誉。

回顾本人从事自然保护事业几十年来的人生经历：既有自然保护技术工作坚实的基层专业基础，也有国家自然保护最高行政管理决策的高层视野；既有中国自然保护事业必须提高全民素质的感受，也有有机结合个人兴趣爱好与工作而引领中国生态摄影事业发展的实践；既有跑过相当数量的国内外自然保护地的难得经历，也有履行国际公约而代表中国与全球自然保护同行相互交流的经验。由于有这三个车轮共同前进的驱动力，无论是退休前，还是退休后，我都甘愿把自己的年华沉浸在祖国广博深邃的大自然中，不知经历了多少次翻山越岭、多少次风餐露宿和多少个不眠之夜，竭力付出了大量精神和体力，有时还不得不侧身站着，抵御暗处的冷言冷语，甚至不端压力。总之，喜怒哀乐皆成生命。

至今，这套丛书能全部出齐，确实不易。在此，我必须衷心地感谢在困难时给予我无私支持和帮助的领导和同事们。感谢原林业部调查规划设计院、北京林业大学、原林业部、原国家林业局、国家濒危物种进出口管理办公室、国家林业和草原局给我的普通人生提供了最好的舞台、最大的支持和最有效的鼓励！感谢各省(自治区、直辖市)林业（和草原）厅（局）及野生动植物保护和各类自然保护地管理部门同仁们的大力支持！感谢诸多各类自然保护地、动（植）物救护中心（园）基层工作人员包括一线护林员给予我的尽可能的帮助！感谢诸多国家及地方科研院所（校）的专家及尊敬的多位院士给我的无私支持和指教！没有你们对中国自然保护事业的默默奉献和保护成就，没有你们对我本人的积极支持和热情帮助，这套丛书是万万不可能经过这么多年的考察、积累，并最终完成资料收集、立意构思和出版发行的。我必须在这里再一次表示衷心的、深深的感谢！最后还要感谢我的家人对我这个不爱着家和在家就读书、埋头整理资料而“不爱理家事”的老头儿的无私支持和帮助！

本书的出版得到了国家林业和草原局唐芳林副局长的热情鼓励和指导，得到了国家林业和草原局草原管理司的大力支持及草原管理司颜国强处长的指导，得到了北京林业大学草原学院董世魁院长、中国科学院植物研究所白永飞研究员、北京大学李文军教授、江西林业科学院黄晓凤副院长等专家的热情指导和帮助，尤其是得到了中国草原泰斗任继周院士极大的肯定与支持并亲自作序。在此，我表示衷心的、特别的感谢！

书稿今天终于完成了，虽然不完全尽人意，但是心血滴滴。书中可能有科学知识把握不准的地方及配图效果不好的地方，也可能存在一些错误和缺陷。在此，我诚恳地希望读者多多提出宝贵的批评意见！谢谢！

陈建伟

2025年4月1日于奥运村行知斋

后记

POSTSCRIPT

The rangelands is a field that is somewhat new yet not totally unfamiliar to me. It is new because it did not previously fall among the mandated responsibilities of the former Ministry of Forestry (or the State Forestry Administration), which had authorities only over what's known as the "three ecosystems and one diversity" — namely the ecosystems of forests, wetlands and deserts, and wildlife biodiversity. It is not altogether an unfamiliar field because, as a person long engaged over the past few decades in surveying and monitoring, researches and education, as well as in administrative jobs related with the "three ecosystems and one diversity", I have inevitably often ventured into the vast rangelands and grasslands of the country when I was responsible for the overall development and management of nature reserves nationwide.

Be it in terms of scientific theories or in practice, there is frequently a sort of mutually interlocking and un-severable relation between the ecosystems of forests, wetlands and deserts and that of rangelands. Take for instance, forests often come side by side with grasslands, with either the former surrounded by the latter, or the latter dotted by the former; whereas the shrubbery land is often taken as a category of forests, there is also an entry among the list of rangeland types known as shrubbery grasslands. Take for another example, places that come with both grasses and water can either be categorized as wetlands or grasslands; take as one further example, while the list of deserts contains such a type as grassland deserts, that of rangelands also contains a type known as desert rangelands. When a given place is designated as a nature reserve, it is often the case that it comes with all the ecosystems of forests, wetlands, deserts as well as that of rangelands. Credited as the most effective means of protecting the holistic ecosystem, is it possible for us to say that, for instance, this is a forest-featuring nature reserve, therefore the protection of wetlands, deserts and rangelands is none of our business? It is more such a case when it comes to the protection of wildlife, given that they often find their shelter, in forests, forage on grasslands, drink from wetlands, and play over deserts. Over the decades when I worked as a practitioner of nature conservation, I have often found myself face to face with rangelands. For the above-mentioned reasons, it is reasonable for me to say that this is not a field that is altogether unfamiliar to me.

During the structural reform launched by the State Council in 2018, the authority over the rangelands and grasslands, which formerly belonged to the Grassland Division affiliated to the Department of Husbandry under the Ministry of Agriculture, was transferred to the National Forestry and Grassland Administration (NFGA) that was newly reorganized on basis of former State Forestry Administration (SFA). In my understanding, this change is by no means a mere minor adjustment in the job delineation among different government agencies. Instead, it marked a fundamental shift in the philosophy of rangeland management, from one that focused purely on their productive functions to one that gives equal attentions to both their productive and ecological functions, and in this sense, as I believe, it symbolized a major elevation in the role of rangelands as a crucial composing element in national ecological security.

To better understand rangelands, I, in addition to all the remedial courses I have taken concerning science-based theoretical knowledges about them, I embarked on manyself-funded field trips to visit the rangelands and grasslands across the country as well as to learn from specialists in this field. And, in order to gain a more objective view concerning rangelands in China as against the global perspective, I also went out of my way to go on field trips to Russia, Mongolia,Kazakhstan, Uzbekistan, Ecuador, Australia, New Zealand and other countries where the rangeland and husbandry industry is relatively better developed.

I, through all these field trips of learning and observation, got keenly aware that the rangeland ecosystem is perhaps the most humble and the most indistinguishable one among all the terrestrial ecosystems (such as that of forests, wetlands, deserts as well as cultivated croplands). They have neither the majestic and easily recognizable masculinity of forests, or the meandering and subtle feministic delicacy of wetlands, nor the awe-inspiring grotesqueness of deserts, or the tenderly-cared meticulousness of cultivated croplands. Quite on the contrary, the tiny herbaceous plants are often trodden beneath people's feet and dismissed off as being worthless. But, grasses are at the same time so great! In spite of their tiny blades and fragile limbs, they, countless of them, cling tightly together to form a thick protective blanket for our planet Earth, mighty and endless as the wave-filled oceans, tender and caring as the soft skin. Being the essential composing element of the largest green vegetation on the land of China, grasses can be ubiquitously found everywhere, in forests, on wetlands, and over deserts. Tucked amidst crevices of dry and barren rocks, they testify to the unyielding vitality of life; Thriving among soil of wet and fertile lands, they present vivid interpretations for the unquenchable perseverance as depicted by a line taken from an ancient Chines poem — Merciless the temper of wild fires might be, grasses would sprout once more as spring breezes return. So tiny and fragile are grasses, yet the vigor contained in them is so powerful. Among the forever circulating cycles of the life system that exists on this planet, grasses are mere primary producers, yet they are also the unyielding great heroes who never shirk from their duties just because of their humble status. Compared with a life constantly under the spotlight, they prefer one that is quiet yet fulfilling. In my judgment, whatever compliments that we human beings are capable of creating, they make up no adequate praises to the great virtues showcased by grasses. We owe our respects to rangelands and have the obligations to look at them with a fresh viewpoint and from a higher level.

As far as the protection of forest ecosystem is concerned, in order to keep forests free from or less subject to human interference, we have established a lot of nature reserves in such forms as no-logging zones, natural forest reserves and so on, hoping that forests kept under this conditions will maintain, or be restored to, their optimal communities where all kinds of plants, animals and microorganisms can live freely in a well-balanced ecological state, thus making it into a system boasting the best ecological state, the ideal internal structure, and the most comprehensive biological chains. It is reasonable to claim that a forest ecosystem that is the least subject to human intervention can be regarded as one that comes with the best ecological functions (with the exception of artificially developed commercial plantations used purely for timber production). We should not only protect the existing natural forests, but also make efforts to cultivate some secondary forests and even plantations, all of which will in turn be developed into ecological forests where top communities can exist and thrive. The reason is because what we human beings need the most from forests — the dominating body of terrestrial ecosystems — is the natural, irreplaceable, and comprehensive ecological functions that they provide, and that they are ecosystems that must be kept in intact state as much as possible.

As to the cropland ecosystem, it is a well-cultivated one where deliberate intervention and attentive cares from people are needed at every stage throughout the process, ranging from land preparation and fertilizing, seed selection and breeding, crop nurturing all the way through to the final harvest. In other words, the cropland ecosystem is one that is by nature filled with intentional human interventions all the way through its entire process; the more meticulous the cares given by people are, the more reliant on science-based human interventions it is, the better the cropland ecosystem it will be. The ultimate goal that human beings wish to achieve through cultivating farmland is to produce more food, thus making such an ecosystem a pure productive one on which human beings depend for survival.

Unlike either the forest ecosystem where the less human intervention there is, the better the situation will be, or the cropland ecosystem where the more intentional human intervention there is, the better the situation will be, the rangeland ecosystem is a symbiosis made up of grasses (including wild plants), livestock (including wild animals) and human beings, one that comes with both productive and ecological functions. For sure, the rangeland ecosystem we are talking about here refers mainly to the animal husbandry-based ecosystem (given that there is now hardly any primitive rangeland that is absolutely free from the impacts of animal husbandry, and that purpose-developed artificial pastures for hay harvesting does not belong to what this book is concerned either).

The rangeland ecosystem at issue here is a dynamically-balanced one in which grasses, livestock and human beings coexist in a mutually influencing and mutually restraining manner. Such an ecosystem won't do when either too much or too little human interventions are exerted on it. The extensive degradation and sandification of rangelands that is ongoing is just the consequence of excessive man-made interventions that have arbitrarily broken the dynamic balance that the system previously had.

The "producers" of rangelands, namely the herbaceous plants, small shrubberies and forbs (hereinafter referred collectively to as "the grasses"), produce nitrogen. Of the nitrogen thus produced, about 2/3 is consumed by the "primary consumers", namely the domesticated livestock like cows, sheep, horses, camels as well as wild animals like antelopes, wild camels and *etc.* (hereinafter collectively referred to as the "livestock"); In addition to the approximately 53% that is digested and absorbed, around 35% is returned to the soil on rangelands in the form of feces and urines. Afterwards, the "secondary consumers" like wolves, foxes, raptors, snakes and *etc.* join the processes of energy conversion through preying on these "primary consumers". The "decomposer", namely microorganisms and certain soil-borne insects, will be the next players to come on stage. They would finish the last step in the conversion, thus making 97% of them be absorbed and utilized by the grasses. In this recycling process of lives, "grasses" make up the primary productive force, "livestock" make up the driving force, whereas "humans" make up the "secondary consumers" as well as the "facilitating force" of the "destructive force". Among the three, "humans" exert the greatest impacts. When the growth of "grasses" and the foraging of "livestock" reaches a state of dynamic balance through the regulation of "humans", the system will facilitate the continuous renewal and sustainable use of the rangeland ecosystem, turning it into a virtuous cycle. Otherwise, the rangeland ecosystem will be damaged, even to the degree of total collapse.

Therefore, a sound understanding about rangelands should neither focus merely on animal husbandry but ignore ecology, nor strive for the so-called "ideal" state of rangelands without taking into consideration the needs of humans and animals. We should never take the number of cows and sheep as the measuring stanrdards for success or failure, nor can we take total absence of cows and sheep as the ideal state. We must clearly recognize that the rangeland ecosystem is a dynamically-balanced one in which grasses, livestock and human beings coexist in a mutually influencing and mutually restraining manner, and that it is a very unique one that is different from forests, wetlands, deserts, farmlands and oceans.

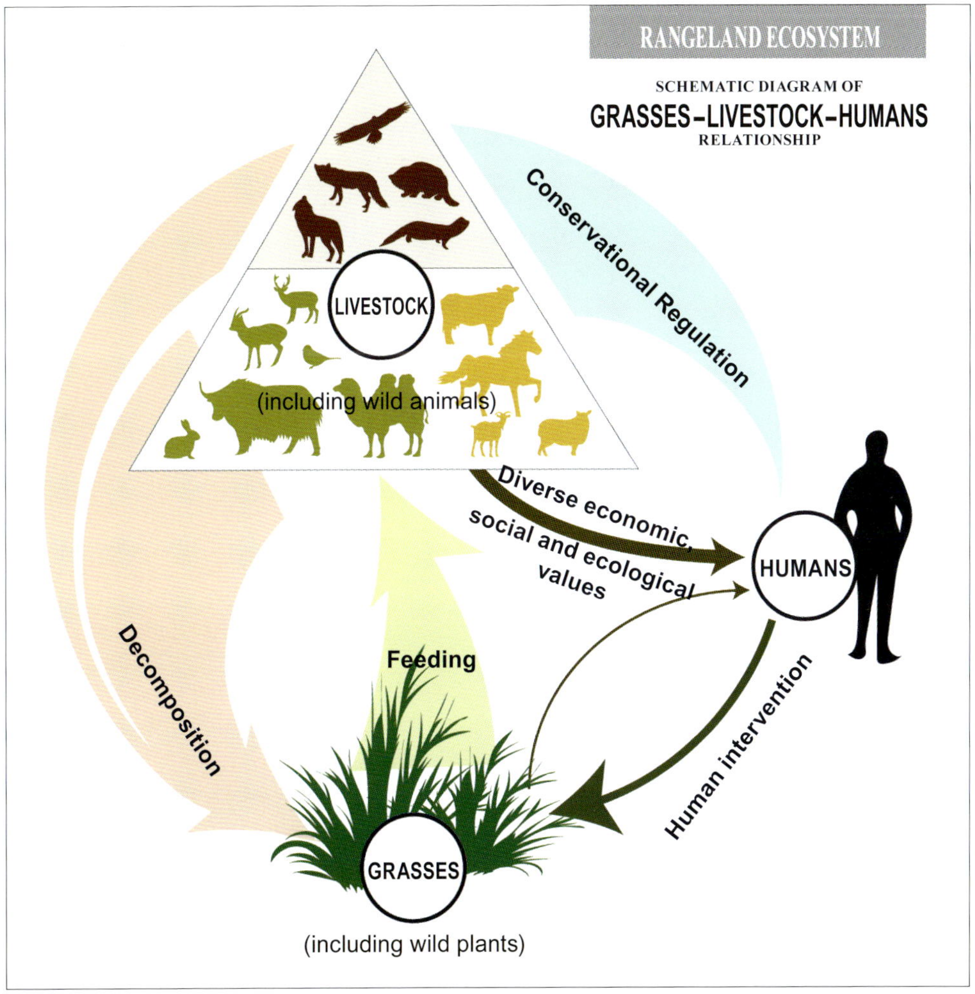

Nevertheless, in spite of their miniature stature, grasses come in a bewilderingly huge variety, not to mention the fact that they are widely distributed in various horizontal, vertical and geographical zones. It is indeed a daunting challenge to tell about the story of the diverse rangelands in China through such a brand-new medium of eco-photography. It not only calls for plain and easily understandable languages to drive home to readers knowledges concerning the complicated, and often confusing, types of rangelands. I also need to select photos that are representative enough to tell about the defining features of each type of rangelands whose features are often not so obviously identifiable, which is a hard problem that I had not encountered when I worked on previous books that focused respectively on such ecosystems as forests, wetlands and deserts. Still more troublesome was the COVID pandemic that struck now and then over the past three years and made one after another of my field trip plans aborted. As a result, I could only content myself by delving into books and various reference resources to search for relevant information, and by combing through the millions of photos that I have previously collected. At the wake of the pandemic, I also seized various opportunities to go on some highly intensive field trips to some rangelands so as to make up for what was still missing.

As someone who has personally witnessed China's modern

cause of natural protection, especially that launched since the reform and opening up policy was implemented, I have always, whether during my working years or after my retirement, been a most enthusiastic advocate for as well as the most committed participant in this grand cause. I take it as my greatest honor and pride to have had this destined opportunity to get involved.

Decades of dedicated work in natural conservation and protection makes me feel keenly that it is far from adequate to rely merely on government-driven efforts and constant appeal of scientists; instead, heartfelt awareness, voluntary involvement and spontaneous actions in natural protection on the part of the general public are all essential for any success in this grand endeavor. It is this realization that inspired my determination to write a series of popular science books on China's various ecosystems, hence contributing my modest shares in enhancing people's natural protection awareness and calling on their active involvement in this field.

Ever since 2007 when I started working on the "A Single Drop of Water Ecological Photography Album series", my determination and efforts have never for a moment faltered. In spite of the misgivings that I had in those early days concerning whether my knowledge storage and capability would be sufficient enough to enable me to get through with the entire series, I wrote down the following to help me pluck up my courage and hold on with the work: "If a single drop of water can play its role in reflecting the glory of the Sun, what about countless drops of them?" So long as I stick to it and don't give up, every time a new book under the series is done, I would be one step closer to the final goal.

"Come whatever may, just keep calm and carry on!" It is this faith that encouraged me to strive dauntlessly forward. In 2008, the first drop of water—*A Single Drop of Water Ecological Photography Album* — was out, followed consecutively by the second — *Diverse Forests of China* in 2011, the third — *Diverse Wetlands of China* in 2014, the fourth — *Diverse Deserts of China* in 2017, the fifth — *We Live Here: Diverse Wild animals in China* in 2021, the sixth — *Panoramic Wetlands* in 2022, and finally the seventh — *Diverse Rangelands of China* that is coming out soon. Up to date, I have not only fulfilled my long-held aspiration to cover "the three ecosystems and one diversity", but am also so fortunate as to live up to the call of the present historic moment when former State Forestry Administration is reorganized as the National Forestry and Grassland Administration and have this last book under the series—*Diverse Rangelands of China* published in 2025.

I feel tremendously delighted that I, over the last 17 years, have successfully completed 7 books that cover all the natural ecosystems as well as wild animals that fall within China's holistic terrestrial ecosystem. Quite a few among these books have been presented by the Chinese government as gifts to international organizations or on important events like Beijing Olympic Games, UN COPs and others. Some have also won prestigious national book awards, the Gold Award for Excellent Pop Science Books conferred by Chinese Federation of Pop Science Writer, the Best Pop Science Books awarded by the Ministry of Science and Technologies, and the Liangxi Pop Science Books Awards being just a few among these honors. Domestic and international media as well as friends from home and abroad have also been highly commendatory about all these books.

Looking back at my decades-long career dedicated to the cause of nature conservation and protection, I am glad that I have had both solid professional expertise fostered through working on nature conservation frontline posts and macro-visions developed through my engagement in policy-making at the nation's highest nature protection administration; that I have been driven both by the objective demands to enhance the general public's awareness concerning nature protection and by my strong personal interest in playing a leading role in pioneering China's eco-photographing cause; that I have had rich experiences both in personally stepping onto protected natural lands at home and abroad and in sharing and exchanging extensively with my global counterparts during the process of fulfilling our obligations to international conventions on behalf of the Chinese government. Motivated by the favorable chances I had in the above mentioned three aspects, I have always, be it during my working years or after my retirement, taken it as a labor of love to venture deep into the vastness of nature in our country, year in and year out, be it rainy or shiny, and regardless of how much painstaking and energy-consuming such journeys might turn out to be. In some cases, I have also had to stand up solitarily against unexpected verbal slings and arrows, even unjustifiable demands and pressure. However, all these — sorrow or pleasure, bliss or curse — are but parts that make life an integrated whole.

It has indeed not come by easily that I finally get all the books under this series published. My heartfelt appreciation goes to all the colleagues and leaders that have offered me their unselfish and unreserved hands whenever I run into difficulties. I owe a sincere thank-you to the Survey, Planning and Designing Institute under the former Ministry of Forestry (MoF), the Beijing Forestry University, the MoF (former), the SFA (former), the Office of Endangered Species Import and Export, and the National Forestry and Grassland Administration, each among which has provided me with a best stage and has been most supportive to me so that I can demonstrate my capabilities and live my ordinary life to the fullest! I'd also like to say thank-you to all my peer fellows working at the provincial forestry and grassland administrative authorities, at the various wildlife protection agencies and at the management of all nature reserves across the country for the tremendous supports they have afforded me! Thank-you to all the frontline workers and forest rangers in the numerous nature reserves, wildlife protection and rescue centers for the helpful hands you have offered to me! Thank-you to all the professionals, specialists working at various national or local research institutions and academies, to the many respectable academicians for the unselfish supports and insightful consultancies that each of you have provided! Hadn't it been for your whole-hearted commitment to China's nature protection cause and the remarkable accomplishments you have achieved in this regard, hadn't it been for all the unreserved supports and heart-warming helps you offered to me myself, it would hardly have been possible for me to bring the ecological photography album series out and put them on the market, particularly when the painstaking, time-consuming work that are involved in the process — the arduous field trips to the wild, the challenges in information retrieval and literature review, the demanding tasks in sorting out all the materials and in actually writing them down in clear and coherent languages — are considered. So let me once again extend my most sincere, most heartfelt gratitude to you all! Last but not least, my innermost gratitude goes to my family, for the unreserved love and considerate cares each of you consistently downpours on me, and for your patience in bearing with this "domestically-good-for-nothing" old guy, who is either constantly away from home, or would be busy, when he does occasionally stay in, with his reading and researches.

While this book is going through the publication pipeline, Tang Fanglin, deputy director of NFGA, has afforded me with his heart-lifting encouragements and thought-inspiring advices. The NFGA Rangeland Administrative Department and Yan Guoqiang, director of the department, have constantly played highly constructive roles in facilitate smooth progress in the process. Other specialists who have also been highly helpful include: Dong Shikui, the Dean of the School of Grassland Science at the Beijing Forestry University; Bai Yongfei, the researcher of the Institute of Plant Researches the Chinese Academy of Sciences; Professor Li Wenjun of Peking University; Huang Xiaofeng, the deputy director of the Forestry Academy of Jiangxi Province. I feel particularly honored that Academician Ren Jizhou, the founder and pioneer of rangeland science in China, has been fairly commendatory of the book and has personally written a foreword for it. I hereby extend my very special and most grateful thanks to each of the above-mentioned persons.

Now that I am finally through with the manuscript of the book, which, though not totally satisfactory, is the outcome of painstaking efforts. I am keenly aware that it is bound to have places where either the knowledge concerned may turn out to be inaccurate enough, or the photo chosen falls less than desirable in effect. There might also be occasional errors or insufficiencies that have evaded my attentions. It is sincerely hoped that readers will point them out, for which I will be most grateful. Thanks a lot indeed.

Chen Jianwei

Xingzhi Study, Beijing Olympic Village
Apr. 1st, 2025